SpringerBriefs in Earth Sciences

SpringerBriefs in Earth Sciences present concise summaries of cutting-edge research and practical applications in all research areas across earth sciences. It publishes peer-reviewed monographs under the editorial supervision of an international advisory board with the aim to publish 8 to 12 weeks after acceptance. Featuring compact volumes of 50 to 125 pages (approx. 20,000–70,000 words), the series covers a range of content from professional to academic such as:

- timely reports of state-of-the art analytical techniques
- bridges between new research results
- snapshots of hot and/or emerging topics
- literature reviews
- in-depth case studies

Briefs will be published as part of Springer's eBook collection, with millions of users worldwide. In addition, Briefs will be available for individual print and electronic purchase. Briefs are characterized by fast, global electronic dissemination, standard publishing contracts, easy-to-use manuscript preparation and formatting guidelines, and expedited production schedules.

Both solicited and unsolicited manuscripts are considered for publication in this series.

More information about this series at http://www.springer.com/series/8897

Subash Chandra Mahala

Geology, Chemistry and Genesis of Thermal Springs of Odisha, India

Springer

Subash Chandra Mahala
Bhubaneswar, Odisha
India

ISSN 2191-5369 ISSN 2191-5377 (electronic)
SpringerBriefs in Earth Sciences
ISBN 978-3-319-90001-8 ISBN 978-3-319-90002-5 (eBook)
https://doi.org/10.1007/978-3-319-90002-5

Library of Congress Control Number: 2018941994

Printed on acid-free paper

This Springer imprint is published by the registered company Springer International Publishing AG part of Springer Nature
The registered company address is: Gewerbestrasse 11, 6330 Cham, Switzerland

Preface

Thermal springs and geysers are the surface manifestations of earth's inner heat content. Heat energy in the form of subsurface steam, hot water, heated rocks and associated minerals and gas products is considered as geothermal energy. This book presents a comprehensive overview of the nature of manifestation and general aspects of thermal springs of Odisha. Detailed description on geological formations and relationship between tectonic settings with thermal spring manifestation is given for better understanding of the geothermal system. Regional geological mapping indicates that the thermal springs of Odisha are mainly confined to crystalline schists and gneissic terrains of Precambrian period. The thermal springs are located within charnockites, khondalites, augen gneiss and mafic granulites belonging to Eastern Ghats Supergroup or quartzites/quartz schists and metapelites of Iron Ore Supergroup or within Vindhyan Supergroup of rocks. The chemical analyses indicate that the spring waters are not identical to one another in their chemical characters. They are categorized mainly under three types, i.e. (i) sodium chloride (NaCl) type, (ii) sodium bicarbonate ($NaHCO_3$) type and (iii) calcium bicarbonate ($CaHCO_3$) type.

Physico-chemical characteristics of water, as well as gas, have been studied and interpreted to ascertain the nature and source of geothermal fluid and to generate a conceptual model on the origin. The thermal springs of Odisha are considered to be of meteoric origin and the heat source may be derived from geothermal gradient, disintegration of radioactive elements or from the exothermic reactions during metamorphism.

Thermal springs of Odisha belong to low enthalpy geothermal category and hence, their utilization potential is restricted to direct or non-electrical application fields mainly for balneology or therapeutic use, space heating, greenhouse cultivation, industrial applications, mineral water bottling, etc.

Rapid depletion of conventional energy resources has demonstrated that there is an urgent need for exploration of resources to meet the ever-increasing demand for energy. Geothermal energy yet largely untapped is a significant non-conventional source of energy which can be used without polluting the environment in the present state of scientific and technical knowledge. Multipurpose utilization of geothermal energy in various countries of the world has attracted the scientific and administrative community in India to assess the potential of geothermal resource of individual hot springs. Various research and academic institutions have started preliminary and, in some cases, detailed investigations of geothermal fields in order to examine the possibility of suitable utilization of the thermal springs of India. This energy is practically perennial, inexhaustible, pollution free and eco-friendly. A holistic approach would certainly bring awareness about the importance of this energy and protection of its quality and quantity would certainly help in conservation and proper management of this resource. In fact, exploration and exploitation of this energy can open up new vistas in the direction of future energy requirements of the nation.

The book will be a useful reference for the earth scientists, planners and researchers in the field of geothermal energy. Besides, as the thermal springs are spectacular sites for tourists, the book can also help the tourists, tourist guides and tourism departments. I have tried my best to acknowledge all sources of information and views of original authors and workers. Though care has been taken, some might have left unintentionally and I welcome such omission, so that they can be rectified in future.

This book is an outcome of my association with Prof. S. Acharya, Former Vice-Chancellor and Head, P.G. Department of Geology, Utkal University, Bhubaneswar, India. I express my deep sense of gratitude to him for his guidance and valuable suggestions. I am very much grateful to Prof. P. P. Singh, Emeritus Professor, P.G. Department of Geology, Utkal University, Bhubaneswar, India for his inspiration for writing this book. I am extremely thankful to Dr. P. C. Naik, my colleague for his academic criticisms which are of great help in structuring this book.

I sincerely thank Dr. N. R. Das and all my well wishers for their moral supports and inspirations.

Bhubaneswar, India Subash Chandra Mahala

Contents

Chapter 1
Introduction

Abstract General aspects of the thermal springs are discussed in this chapter. They are natural phenomena. Different terminology has been assigned to the thermal springs depending upon the nature of manifestation. A thermal spring is considered as one of the non-conventional source of energy. Exploration and utilization of this source has become vital to achieve better economy in the long run for the country. In fact the geothermal energy is preferred because it is perennial, inexhaustible and pollution free. Thermal springs are documented throughout the world. They are usually observed in and around tectonic zones and often associated with volcanic activity. In India, thermal springs are observed around specific tectonic system and predominantly traced in Himalayan mountain ranges and have been grouped into specific geothermal provinces. The thermal springs of Odisha are located in the tectonic zones such as (i) Mahanadi graben and (ii) Precambrian mega lineaments. These springs are not directly related to any igneous activity. Description on thermal springs of India in general and Odisha in particular are available in published literature. Present work includes geological, hydrological and geochemical characteristics of all the eight thermal springs of Odisha. Further discussions on their manifold uses and a genetic model have also been attempted.

1.1 General Introduction

Thermal spring is a natural phenomenon. It is variously named such as hot springs, mineral springs, magic water, geysers, fumaroles etc. depending on the nature, characters and modes of manifestation on the earth surface. Surface and ground water reaching great depths in a geothermal basin, get heated and partly get converted into steam. The steam-charged hot water reappears on the earth's surface as hot springs and geysers. They are naturally spectacular events as they are sporadic on the surface of the crust expressing earth's internal energy. The geothermal energy is one such source of energy, which is catching the imagination of all, because it is in the form of heat that can yield warmth and power. In other words geothermal energy may be

S. C. Mahala, *Geology, Chemistry and Genesis of Thermal Springs of Odisha, India*, SpringerBriefs in Earth Sciences, https://doi.org/10.1007/978-3-319-90002-5_1

simply stated as the immense storehouse of energy below the surface of the crust created by the mammoth heat engine producing hot spots.

The geothermal energy generally includes thermal energy in the form of subsurface steam, hot water, heated rocks and associated minerals and gas products that can be extracted from hot water and steam phases. Sometimes, the temperature is high enough to convert the descending water into steam with high pressure in form of fountain/jet on the surface as a geyser. Thermal springs, geysers, fumaroles and moffettes are the surface manifestations of earth's inner heat content. Some of the hot water that travels back fast through faults and fractures reaches the earth surface prior to getting cold. Albeit some of it stays deep underground, trapped in cracks and porous rocks and the natural collection of hot water is called geothermal reservoir. The axiom of internal constituent has demonstrated that vast quantity of heat is stored in the lower layers of the earth's crust. The lava flows carry heat from the earth's interior. At certain times the magma remains below the crust, heating nearby rock and water (meteoric water that has seeped into the earth).

Recently it has been found necessary to concentrate on finding and tapping of alternative/non-conventional sources of energy for sustainable socio-economic development. It is known that economic development of the country depends on consumption of energy derived from fossil fuel. This has brought in futuristic ecological catastrophe by increase of carbon in the atmosphere, leading to greenhouse effect and global-warming. There are vast reservoirs of untapped energy, which can possibly produce electricity and heat useful to humans and industries without burning fossil fuel. The future generations have the right to ask for a better living in terms of lesser intensity of global warming.

Energy is the basic need to the national economy of a country. A direct consequence of the energy crisis has been the dilemma/setback of many nations. Countries not adequately endowed with fossil fuel or hydel resource, are searching for alternative sources to meet the ever-spiraling energy demands. Rapid depletion of conventional energy resources has demonstrated that there is an urgent need for exploration of resources to meet the ever-increasing demand of energy. Thermal springs, as the significant non-conventional source of energy, are yet largely untapped in the present state of scientific and technical knowledge. This is also considered as the most suitable non-conventional source of energy, which can be used without polluting the environment.

The recent increase in utilization of geothermal resources is due to number of factors including improvement in technology, environmental advantages and economic aspects. Technological improvements include the spectrum of activities for initial exploration resulting in the production of electrical power. Geothermal energy has the advantage that it does not depend on seasonal variations (hydro-energy) and, therefore, it is possible to maintain a fixed constant level of energy production throughout the year. From the environmental standpoint geothermal power production is benign compared with the burning of the fossil fuels. Hence, it is likely to be economical as compared to conventional energy sources. Thus, geothermal energy utilization can significantly reduce effect of green house gases and acid rain.

The energy sector in India configures identification of resources, imbibes new technology, and plans to avoid energy crisis and assurance of environmental safeguard. Multipurpose utilization of geothermal energy in various countries of the world has attracted the scientific and administrative community in India to assess the potential of geothermal resource of individual springs. National Geophysical Research Institute (NGRI), Geological Survey of India (GSI), Variable Energy Cyclotron Centre (VECC) and various other academic institutions have started preliminary and in some cases detailed investigations of geothermal fields in order to examine the possibility of suitable utilization of the thermal springs of the country.

The geothermal energy can either be used by utilizing the steam to produce electric current in power station or be applied directly as primary heat source. Application of this potent, economical and viable source of geothermal energy will not only reduce and conserve conventional energy (coal, oil and natural gas) resources but also minimize environmental degradation.

1.2 Nomenclature

In the simplest sense, thermal springs are the springs that issue water at temperatures substantially higher than the atmospheric temperature of the surrounding and come from great depth. These are considered to be natural phenomena expressing earth's internal energy. In a geothermal field, water trapped in fractures in the near surface rocks is subsequently heated by the sub-crustal heat. The natural geothermal gradient, decay of radioactive materials and exothermic reactions have influenced the heating process. Later on the hot water and steam escape to the surface of the earth along the fractures. There is a subtle distinction between the terms thermal spring and hot spring. A thermal spring can be called a 'hot spring' when the temperature of water reaches the boiling point (100 °C). The hot water of these springs is charged with several mineral matters because of its ability to dissolve them (mineral matters) from rocks while moving through them due to high temperature. Thus, the hot springs are also known as "mineral springs". The spring charged with minerals impart curable property of the water and named 'magic water'. The tradition-bound people in India often consider hot springs as holy pilgrim spots because of their special balneotherapic characters.

1.3 Classification of Geothermal Systems

Geothermal systems are classified by a series of descriptive parameters such as reservoir equilibrium state, reservoir fluid type, reservoir temperature, host rock and heat source etc.

A. Reservoir equilibrium state: This is the fundamental division of geothermal system and is based on the circulation of reservoir fluid as well as the mechanism of heat transfer. Based on these parameters the geothermal systems are categorized as (i) dynamic systems and (ii) static systems. Dynamic systems are continually recharged by water entering the reservoir, heated up and then poured out of the reservoir either to the surface or underground horizons. Heat is transferred through the system by convection and circulation of fluid. Static geothermal systems have only minor or no recharge of water to the reservoir and heat is transferred only by conduction.
B. Fluid Type: Geothermal reservoir fluid state consists mainly of water (liquid dominated) or steam (vapour dominated). In most cases both steam and liquid water co-exist in varying proportions. Based on this factor, the geothermal systems may be classified as springs, geysers or fumaroles.
C. Reservoir Temperature: Based on the temperature or enthalpy of the reservoir, they are commonly described as low temperature (<150 °C) or high temperature (>150 °C) geothermal systems. Nevertheless the discriminating temperatures are not rigid hence, some workers use the term intermediate type to indicate reservoir temperatures between 120 and 180 °C.
D. Host Rock: Based on the rock types containing geothermal reservoir, the system, is classified as sedimentary, metamorphic and volcanic systems.
E. Heat source: Heat source is a function of geological phenomenon or tectonic setting. If the heat flux is provided by magma, the system is called volcanogenic and they are invariably high temperature systems. If heat is supplied by tectonic activity or deep circulation of water due to geothermal gradient, then they are called non-volcanogenic systems that include both low and high temperature geothermal systems.

McNitt (1970) after reviewing various attempts in classification of geothermal systems proposed that the geological processes include the position of field with respect to orogenic and volcanic belts. According to him various geothermal systems are conveniently classified into two major types:

1. Cyclic geothermal system and
2. Storage geothermal system.

Cyclic Geothermal system: Here the hot water is meteoric water, which has passed through a cycle of deep descent, heating and rising by convective forces. Cyclic hydrothermal systems reflect the general tendency for meteoric water to circulate down through the rocks to depths controlled by the local geological structure. They are further sub-divided as follows-

- High temperature systems associated with recent volcanism
- High temperature systems in non-volcanic regions of tectonic activity
- Warm water systems in near normal heat flow zones.

Storage Geothermal systems: In this system the water is stored in rocks for long period geologically and heated up insitu. They are further sub-divided as follows- (1) Sedimentary basin systems and (2) Metamorphic systems.

The geothermal systems of India are of medium enthalpy nature and can be broadly grouped into the following categories (Shanker et al. 1991).

(a) Those associated with younger granites as in Himalayas
(b) Those associated with major tectonic features or lineaments as in the west coast areas of Maharastra
(c) Those associated with rifts and grabens of Gondwana basins of Damodar, Godavery and Mahanadi valley
(d) Those associated with thick quaternary and tertiary sediments in the Cambay basin of West Coast.

1.4 Geothermal Provinces

Geothermal provinces are found throughout the world in a range of geological settings and are increasingly being developed as a non-conventional energy source. Each of different types of geothermal systems has distinct characteristics that are reflected in the chemistry of geothermal fluids. Geothermal provinces are commonly classified or divided by a number of descriptive terms such as liquid or vapour dominated, low or high temperature, sedimentary or volcanic hosted etc.

1.4.1 Geothermal Provinces of India

Based on the studies carried out in different parts of the world, it is now considered that presence of an active heat source either cooling magmatic body or recent volcanism and channels permitting up-flow of thermal fluids from deeper levels to the surface are generally responsible for the surface geothermal manifestations. The occurrences of geothermal energy manifestation (more than 300 in number, Shanker et al. 1991) are widely distributed in Peninsular and extra Peninsular India. The temperature of these springs varies from just over 5 °C in excess of the mean ambient temperature of the area to the boiling point of water. Deb (1964) mentioned four important belts where the maximum numbers of thermal springs are known to occur. He further mentioned that, beside these main occurrences, there are number of thermal springs along the Mahanadi valley of Orissa. Gupta (1974) classified the thermal springs of India into four groups on the basis of tectonic movements. These are:

(i) The Himalayan mobile Belt
(ii) The West coast region
(iii) Narmada-Son rift zone and
(iv) The hot springs distributed in the eastern part of Peninsular India, in Bihar and Bengal states in two major belts, the northern Rajgir-Monghyr belt and the southern belt more or less parallel to the boundary faults of Gondwana field.

According to Singh and Bandyopadhyay (1995) Indian geothermal fields are distributed over generalized geological base with a simplified stratigraphic legend such as Archaean-Proterozoic sequence (including both meta-sedimentary and magmatic rocks), Palaeozoic–Mesozoic sequences (comprising marine and continental deposits) and Tertiary-Quarternary sedimentary formations. The occurrence of geothermal fields in India relates to:

I. Tertiary and quaternary magmatism or metamorphism in Himalayas.
II. Mesozoic and Tertiary block faulting and epiorogenic activity in shield areas with no apparent evidences of late tertiary or younger magmatism.
III. Zones of neo-tectonism or recent seismicity.

Krishnaswamy and Shanker (1982) mentioned twelve geothermal provinces of India based on the geo-tectonic setting and their distribution. The occurrences of these thermal areas are controlled by basic structural and tectonic framework of peninsular and extra peninsular India. On the basis of available data ten well-defined geothermal provinces could be recognized in India by considering their geographical location, geo-tectonic setting and geothermal characteristics (Ravishanker et al. 1991) (Fig. 1.1). These are:

a. Himalayan geothermal province
b. Naga-Lusai geothermal province
c. Andaman-Nicobar island geothermal province
d. West coast geothermal province
e. Cambay graben geothermal province
f. Aravalli geothermal province
g. Son-Narmada-Tapi geothermal province
h. Godavari valley geothermal province
i. Mahanadi valley geothermal province
j. South Indian cratonic province.

1.4.2 Geothermal Provinces of Odisha

The thermal springs of Odisha in particular can be pigeonholed into the following categories of geothermal environments:

(i) Mahanadi valley geothermal province—Boundary fault along Gondwana graben
(ii) Pre-cambrian geothermal province—NNE-SSW lineaments in Precambrian terrain.

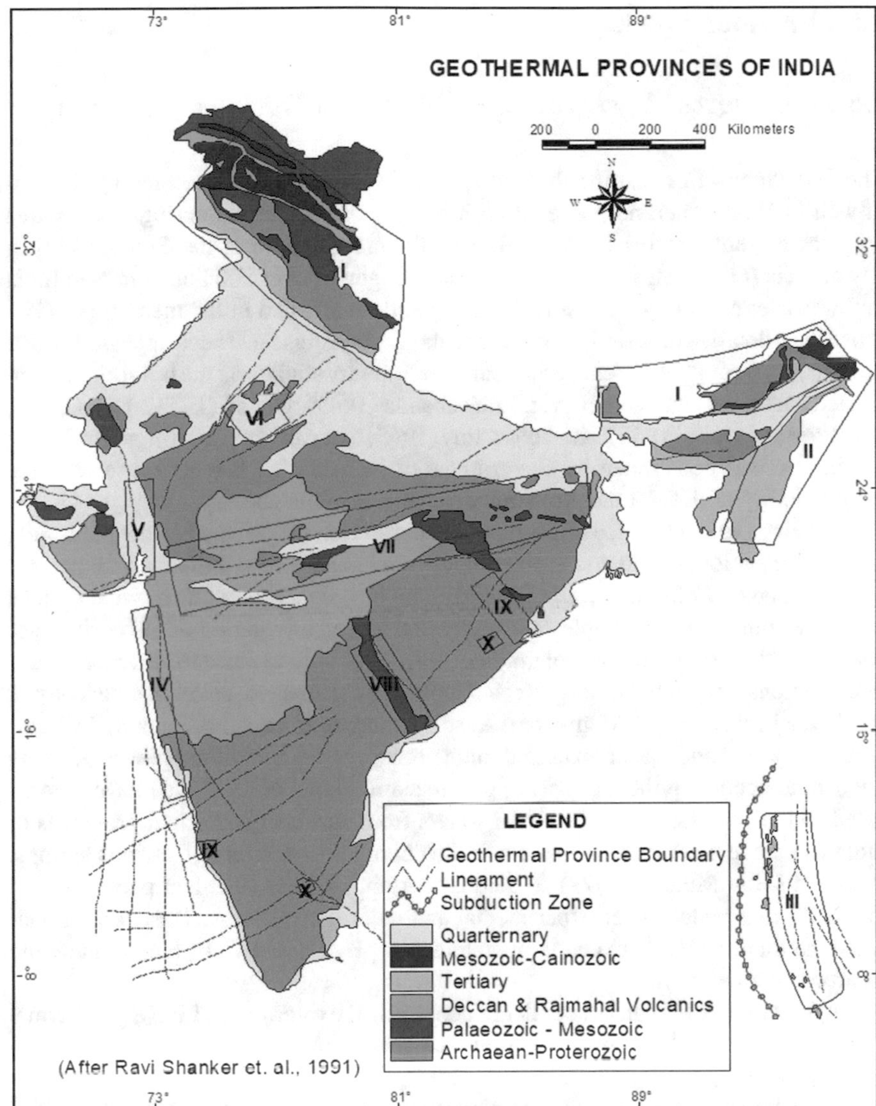

Fig. 1.1 Geothermal provinces of India

According to Ravishanker et al. (1991) the Mahanadi valley geothermal provinces of Odisha are categorized as of Group IX (Fig. 1.1). Majority of thermal springs of these belts tend to align themselves either along or parallel to the major fault zones or along their associated faults. Acharya (1966) invokes the interrelationship between the Gondwana tectonics (along Mahanadi valley) and the occurrence of thermal springs.

1.5 Previous Work

1.5.1 Historical Background and Previous Work

The first attempt to enlist the thermal springs in India was documented by Schlagintweit in 1862, when he prepared an inventory of 99 thermal springs in the area covering present territorial limits. Oldham, the first Director of the Geological Survey of India (GSI) prepared an exhaustive catalogue of over 300 hot springs in India that was later edited by his son and subsequently published in the memoirs of GSI. Since then, detailed descriptions of several thermal springs have been appeared which lie scattered through different periodicals. Preliminary studies on the hot springs were carried out by various workers viz. Pranvananda (1949), Ghosh (1954), Deb (1964), Chaterjee and Guha (1968), Krishnaswamy (1965) and Acharya (1966). A Hot Spring Committee was constituted by Government of India during late sixties to examine the possibility of development of some hot springs in the country.

Realizing the importance of geothermal energy Deb (1964) made a detailed study on the distribution, origin and utilization of waters and gases of the thermal springs of India. Deb and Mukherjee (1969) worked on the genesis of a few thermal springs in Chotnagpur gneissic complex and suggested a deep circulation of meteoric water heated due to normal geothermal gradient aided probably to some extent by exothermic reactions and radiogenic sources. Gupta (1974) made a detail account on the geothermal resources of Manikaran, Kasol, Khirganga, Vasistha, Puga and Chumtang hot springs and has emphasized on the prospects of generating power by using the natural occurring thermal spring water/steam. Gupta et al. (1975) have investigated on the geochemistry of thermal waters from various geothermal provinces of India and made a comparative study on the chemical characters of various springs. Saha and Datta Munshi (1979) studied on certain physico chemical parameters of thermal spring water under experimental and natural conditions. They further analyzed the relationship between temperature and pH, pH and CO_2, temperature and bicarbonate alkalinity.

Gupta et al. (1979) carried out detail geophysical exploration of Puga geothermal field and assessed the power potential. Krishnaswamy and Shanker (1982) have done a pioneering work on the scope of development, exploration and assessment of geothermal resource potential of the country in detail. Saxena and D'Amore (1984) investigated on the aquifer chemistry of Puga and Chumtang geothermal systems and suggested that Puga area is quite promising for electric power generation. Datta Munshi et al. (1984) analyzed the gases from the thermal springs of Bihar and reveal the presence of rare gases like helium and argon in the springs.

Saxena and Gupta (1985) made a detail study on the aquifer chemistry of thermal water in Godavari valley and suggested that the thermal waters of the valley belong to two separate systems. Mukherjee (1985) investigating on the geochemistry and hydrology of thermal springs in Monghyr district of Bihar, proposed a conceptual model on the thermal spring activity in the area. Saxena (1987) studied on the application of gas and water chemistry of various geothermal systems to evaluate sub-surface temperature. Saxena and Gupta (1987) made a geochemical study of Konkan coast geothermal waters to evaluate sub-surface temperatures and utilization potential. Bandyopadhyay and Nag (1989) studied on the hydrology and chemical characteristics of Bakreswar group of thermal springs in Birbhum district of West-Bengal and suggested a meteoric origin of the thermal water, the chemical constituents being derived from rock/water interaction at an augmented temperature.

Regional tectonism and tectonic features have played an important role in the exploration of geothermal domains and delineation of heat flow zones. Gopalkrishnan and Vardan (1996) provide a generalized account of geothermal manifestation of Tamilnadu-Pondichery coast and their possible relation with major tectonic features of this region. Pandey and Srivastav (1996) throw light on chemical interaction and toxicological hazards associated with Indian geothermal effluents.

Prasad (1996) while working on the geothermal energy resources of Bihar, mentioned in a detail the tectonic framework, geochemistry, radioactivity and scope of utilization of geothermal resource of the state. Gupta (1996) studied on the hydrogeology and hydro-geochemistry of Beas valley geothermal system, Kulu district, Himachal Pradesh and suggested that the thermal waters are mainly of Na–Cl–HCO_3 type. He further suggested a mixed type derived from mixing of this Ca–Mg–HCO_3 type and Na–Cl type that is probably derived through water rock interaction. Nagar et al. (1996) suggested that geological, geophysical and geochemical investigations in Bakreswar-Tantloi thermal field and suggested that the tectonic set up responsible for helium and associated gases in the thermal fluids and soil. Muthuraman and Padhi (1996) made a comparative study on the geochemistry of coastal thermal waters of western Maharastra and Tapi basin of central India. According to them even though the thermal waters of the two areas are located within the basaltic flows of Decan traps, these are distinctly different in major ion ratios and their saturation levels are also different indicating the broad independence of thermal springs from being chemically controlled by surface associated rocks.

Saxena (1996a, b) studied the mineral fluid interaction and chemical equilibria of the prominent geothermal systems of India. He plotted the relevant activities of different thermal waters in the theoretical diagram and these diagrams are useful for the study of potential of the geothermal system. Sharma et al. (1996a, b) studied on the prospect of Tatapani geothermal fluid in MP. They deal with various parameters that make this geothermal field suitable to sustain 300 kw experimental power generation unit. Sharma et al. (1996a, b) carried out the environmental isotope studies in Badrinath and Tapoban geothermal areas. Srivastav et al. (1996) studied on the application of gas chemistry in evaluating reservoir temperature of some Indian geothermal systems. They have collected data on gas chemistry from almost all the geothermal areas of the country and made an attempt to estimate the reservoir temperature of eight major geothermal areas of the country using an empirical gas geothermometry. Chandrasekharan et al. (1996) worked on the geothermal energy potential and direct use of geothermal springs of Godavari valley in AP. According to them the geothermal field is capable of generating considerable amount of electric power and can directly be used for developing a variety of small-scale industries.

1.5.2 *Odisha Scenario*

The thermal springs of Odisha were first described in "Jungle life of India" by Ball (1876). He mentioned the occurrence of thermal springs at Attri and Taptapani area. Oldham and Oldham (1882), Ghosh (1954) provided reference on a few of them while reviewing on the thermal springs of India, which included valuable information concerning the occurrences of thermal springs of India. The thermal springs of Odisha has been studied systematically by Acharya (1966). However, Banerjee (1987) has done some work on Taptapani hot spring area in Ganjam district. Deb (1964) reported regarding distribution of thermal springs in India and mentioned about thermal spring occurrences in Mahanadi valley of Odisha. Geological Survey of India (GSI) has published a special publication "Geothermal Atlas of India" in 1991. The report provides information in some detail about the important hot springs of Odisha.

1.6 Present Work

Thermal springs are considered as surface manifestation of hidden energy. In other words it is accepted as the viable non-conventional source of energy. Today in energy consuming world, there is a need of alternative source of energy. Geothermal energy is considered as inexhaustible, pollution free and least cost to the environment. Both the water and heat source of thermal springs can cater to the present demand of energy. It is important to study the physico-chemical characters of water and gas, and nature of heat. Hence, present investigation will assist to acquire knowledge on the character and potential of thermal springs of Odisha. Present work encompasses locating some new thermal springs in addition to already reported in the literature. Detail investigations on hydrological properties and chemical characteristics of the thermal water have been carried out on all the thermal springs. The results have been documented, analyzed and nature of water has been interpreted. Thermal springs are associated with emanation of certain gases including noble gases like helium. On the basis of above findings, attempt has been made to infer the source, nature and characters of the water and heat. The investigations also deal with the study of characters of litho-assemblages and their relationship in and around the thermal spring regions. Attempt has been made to generate a conceptual model on the origin of thermal springs of Odisha.

References

Acharya S (1966) The thermal springs of Orissa. The explorer, pp 29–35

Ball V (1876) Jungle life in India. THOS De La Rue & Co. Bunhill Row, London, p 720

Bandyopadhyay G, Nag SK (1989) A study on hydrology and chemical characteristics of Bakreswar group of thermal springs, Birbhum district, West Bengal. Ind J Geol 61(1):20–29

Banerjee PK (1987) A short note on Taptapani hot spring, Ganjam district, Orissa. Geol Surv Ind Rec 115:1–5

Chandrasekharan D, Rao VG, Jayaprakash SJ (1996) Geothermal energy potential and direct use of geothermal springs, Godavary valley, Andhra Pradesh. GSI Spl Publ No- 45, pp 155–161

Chaterjee GC, Guha SK (1968) The problem of origin of high temperature springs of India. Proc Symp (II), 23rd Int Geol Cong 17

Datta Munshi JS, Billgrami SK, Verma PK, Datta Munshi J, Yadav RN (1984) Gases from thermal springs of Bihar, India. Nat Acad "Sci Lett" India 7(9):291–293

Deb S (1964) Investigation of thermal springs for the possibility of harnessing geothermal energy. Sci Cult 30(5):217–221

Deb S, Mukherjee AL (1969) On the genesis of few groups of thermal springs in Chhotnagpur gneissic complex, India. J Geochem Soc India 4:1–9

Ghosh PK (1954) Mineral springs of India. Rec Geol Surv India 80:541–558

Gopalkrishnan K, Vardan GN (1996) Geothermal manifestation and their relation to geotectonic features along Tamilnadu-Pondichery coast. GSI Spl Publ No-45, pp 25–37

Gupta ML (1974) Geothermal resources of some Himalayan hot spring areas. Himalayan Geol 4:492–515

Gupta ML, Hari Narain, Saxena VK (1975) Geochemistry of thermal waters from various geothermal provinces of India. Publ No 119 of the Int Assoc Hydrol Sci pp 47–58

Gupta ML, Singh SB, Drolia RK (1979) Geophysical exploration and assessment of power potential of Puga geothermal field. Geoviews VI. No-1–4

Gupta GK (1996) Hydrogeology and hydrogeochemistry of Beas valley geothermal system, Kulu district, Himachal Pradesh. GSI Spl Publ No-45, pp 249–256

Krishnaswamy VS (1965) On the utilization of geothermal stream and the prospects of developing the hot springs in the North-western Himalayas. Indian Geohydrology 1(1):27–45

Krishnaswamy VS, Shanker R (1982) Scope of development, exploration and preliminary assessment of the geothermal resource potentials of India. Rec Geol Surv India III, pt 2:18–38

McNitt J (1970) Geothermics (Special issue 2) 1,24

Mukherjee AL (1985) Geochemistry and hydrology of thermal springs of Monghyr district, Bihar, India. In: Geothermics, thermal-mineral waters and hydrogeology, Theofrastus publications, S. A., Athens. pp 151–163

Muthuraman K, Padhi RN (1996) Hydrogeochemistry of coastal thermal waters of western Maharastra and Tapi basin of Central India—A comparative study. GSI Spl Publ No-45, pp 341–347

Nagar RK, Viswanathan G, Sagar S, Sankarnarayanan A (1996) Geological, geophysical and geochemical investigation in Bakreshwar-Tantoli thermal field, Birbhum and Santalpraganas districts, West Bengal and Bihar, India. GSI Spl Publ N0-45, pp 349–360

Oldham T, Oldham RD (1882) The thermal springs of India. GSI Mem No-19, pp 99–161

Pandey SN, Srivastav GC (1996) Environmental hazards of Indian geothermal fields. GSI Spl Publ No-45, pp 375–378

Pranavananda S (1949) Kailash Manosaravar. Quoted in Krishnaswamy VS, 1965

Prasad JM (1996) Geothermal energy resources of Bihar. GSI Spl Publ No-45, pp 99–117

Shanker R, Guha SK, Seth SK, Ghosh A, Ghosh S, Nandy DR, Jangi BL, Muthuraman K (1991) Geothermal Atlas of India. GSI Spl Publ No-19

Saha SK, Dutta Munshi JS (1979) Interactions of certain physico-chemical factors of the hot springs of Bhimbandha, Bihar (India). Biol Bull India 1(3):13–18

Saxena VK, D'Amore F (1984) Aquifer chemistry of Puga and Chumatang high temperature geothermal systems in India. J Volcanol Geoth Res 21:333–346

Saxena VK, Gupta ML (1985) Aquifer chemistry of thermal waters of the Godavari valley, India. J Volcanol Geoth Res 25:181–191

Saxena VK (1987) Application of gas and water chemistry to various geothermal systems of India. J Geol Soc India 29:510–517

Saxena VK, Gupta ML (1987) Evaluation of reservoir temperatures and local utilization of geothermal waters of Konkan coast, India. J Volcanol Geoth Res 33:337–342

Saxena VK (1996a) Hydrogeochemistry of geothermal waters, Konkan coast, India. Gondwana Geol Mag, 2:467–473

Saxena VK (1996b) Mineral fluid interaction and chemical equilibria of prominent geothermal systems of India. GSI Spl Publ No-45 pp 297–309

Schiagintweit R De (1862) Enumeration of the hot springs in India and high Asia. JASB XXXIII:49–58

Sharma S, Nair AR, Kulkarni UP, Navada SV, Pitale UL (1996) Role of geophysics in deciphering geothermal channel in Tatapani area, Rajauri district, J&K. GSI Spl Publ No- 45, pp 233–242

Sharma S, Nair AR, Kulkarni UP, Navada SV, Sharma SC (1996) Environmental isotope studies in Badrinath and Tapoban geothermal areas, Chamoli district, Uttar Pradesh. GSI Spl Publ No-45, pp 223–231

Singh R, Bandyopadhyay AK (1995) Geochemical studies of some thermal springs in Hazaribagh district, Bihar, India. Indian Miner 49(1 & 2):55–60

Srivastav GC, Srivastav R, Absar A (1996) Application of gas chemistry in evaluating reservoir temperature of some Indian geothermal systems. GSI Spl Publ No-45, pp 83–86

Chapter 2
Study Area

Abstract The thermal springs are located at eight different places in Odisha which can be approached by road from the state capital Bhubaneswar. The thermal springs ooze out from a single spout at Attri, Taptapani, Magarmuhan, Bankhol and Boden areas but oozing at several spots clustered together at Tarabalo, Deuljhori and Badaberena. The highest number of spots has been traced at Tarabalo thermal spring area. There is continuous seepage of hot water in and around the spots and the area has been waterlogged and swampy. There are luxuriant growth of screw pines around the thermal spring particularly noted at Tarabalo and Deuljhori. In most of the cases the thermal spring spots are protected by masonry structures. Encrustation of white and yellowish material is marked on the surface as well as parapet wall. Temples are constructed at certain period of times near these springs, as people find them as place of worship because of natural curiosities. Location and surface mode of manifestations are captured through photographs. The physiographic set-up and climatic condition around thermal spring regions are also noted.

2.1 Introduction

Manifestations of geothermal energy in the form of thermal springs are known at eight places in Odisha. These are distributed in Khurda, Angul, Ganjam and Nuapada districts (Table 2.1). The springs are widely distributed covering Easternghat mobile belt, North Odisha craton and western Odisha Proterozoic sedimentary basins. The locations of the springs are shown in (Fig. 2.1).

S. C. Mahala, *Geology, Chemistry and Genesis of Thermal Springs of Odisha, India*, SpringerBriefs in Earth Sciences, https://doi.org/10.1007/978-3-319-90002-5_2

Table 2.1 Location of thermal springs of Odisha

Sl. No.	Name of the thermal spring	District	Toposheet number
1	Attri	Khurda	73 H/12
2	Tarabalo	Khurda	73 H/8
3	Deuljhori	Angul	73 D/6
4	Magarmuhan	Angul	73 G/8
5	Bankhol	Angul	73 G/8
6	Taptapani	Ganjam	74 A/7
7	Badaberena	Angul	73 C/16
8	Boden	Nuapada	64 L/11

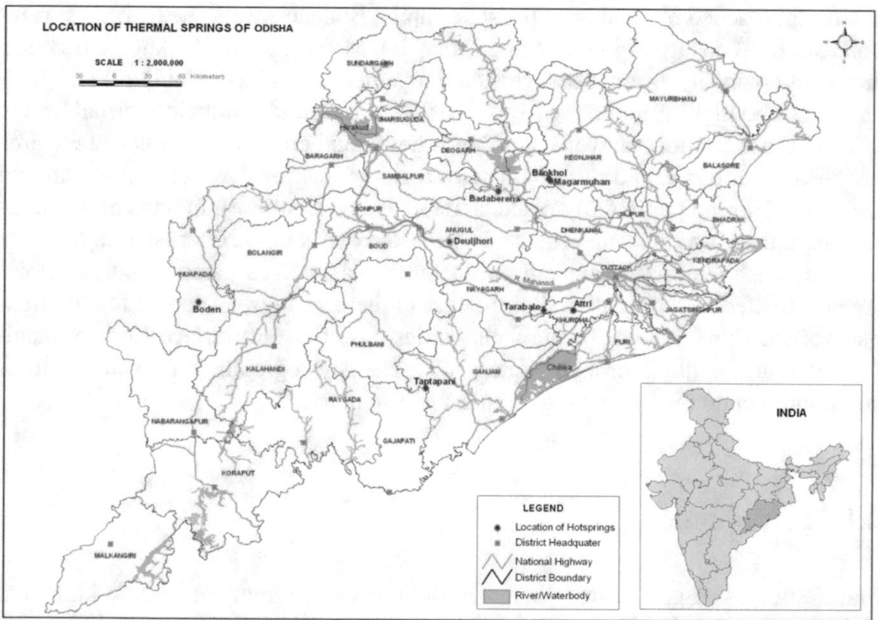

Fig. 2.1 Location of thermal springs in Odisha

2.2 Location and Accessibility

2.2.1 Attri

The thermal spring at Attri in Begunia block of Khurda district is well known since long. Situated at a distance of about 3 km from Baghamari, it is about 48 km away from the state capital Bhubaneswar. Brander mentioned a hot spring at Ooter of Puri, i.e. present Attri (cited by Acharya 1966). In the proximity there is a temple of Lord Shiva, which suggests long connection of the spring with human civilization. Of late the place is developed as a tourist spot by the Tourism department, Government of Odisha and is well connected by all weather black top roads.

2.2.2 *Tarabalo*

Cluster of thermal springs occur near Tarabalo village in Bolgarh block of Khurda district are well known since long. Tarabalo is situated at about 10 km away from Rajsunakhala that is situated also on the State Highway connecting Bhubaneswar and Phulbani. It is located by the side of the Rajsunakhala-Bhapur road at a distance of about 65 km away from the state capital Bhubaneswar. The place is well connected by all weather road and presently being developed as a tourist spot by the Government of Odisha.

2.2.3 *Deuljhori*

The thermal springs at Deuljhori are situated near village Otosinga in Athamallik block of Angul district. They occur at a distance of about 4 km from Athamallik town and are about 250 km away from Bhubaneswar. Hot water ooze out from several spots and each spot is protected by a masonry structure.

2.2.4 *Magarmuhan and Bankhol*

The thermal spring at Magarmuhan in Pallaharha block of Angul district is situated at a distance of about 8 km from Bimla, a small place off Jajpur-Keonjhar Road-Talcher State Highway. Bankhol thermal spring is situated 4 km northwest of Magarmuhan. Magarmuhan is so named as the hot water comes from a rock cutting looking like a mouth of a crocodile.

2.2.5 *Taptapani*

The thermal spring at Taptapani in Ganjam district is situated by the roadside joining Berhampur to Rayagada at a distance of about 55 km away from Berhampur town. The place has also been developed as a tourist centre by Tourism department, Government of Odisha with facility of natural warm water bath. The thermal spring is situated on the Taptapani ghat section and manifest from a single point.

2.2.6 *Badaberena*

The thermal spring near Badaberena village is situated in Chhendipada block of Angul district. The village is well connected by blacktop road with the block head-quarter and the spring site is connected by a metalled village road.

2.2.7 *Boden*

The thermal spring near Boden in western Odisha locally known as "Patal Ganga" is located at a distance of about 4 km away from village Boden in Nuapada district. The place is connected by a metalled road and water comes from a single spot.

2.3 Climate and Rainfall

The state of Odisha enjoys a sub-tropical monsoon climate with three distinct seasons such as summer, rain and winter. The rainy period lasts from mid- June to end of September and receives about 80–85% of the annual rainfall from southwest monsoon. Maximum rainfall occurs between July and August. Normal annual rainfall of the state is around 1482 mm.

The summer is hot with dry temperature rising up to 45 °C and occasionally 50 °C in localized places. During winter season mornings and nights are cool. In general both summer and winter are intense in western and central hilly parts in comparison to eastern coastal regions. The average day temperature during winter is about 22 °C in the eastern part and 18 °C in the western part. The night temperature is about 18 °C in the eastern part and 13 °C in the western part respectively. Extreme heat is felt (50 °C during daytime) in the west where as in the eastern part (45 °C) is felt with high moisture content in the air because of the proximity of the sea.

2.4 Physiography

The state may be broadly divided into three major geomorphological units according to Acharya (2000). They are

(a) The Eastern alluvial plains (coastal),
(b) The Central plateau: (i) The northern plateau (IOG) (ii) Southern plateau (Easternghats) (iii) Mahanadi basin graben and (iv) Intermountain valleys
(c) Western plateau.

The coastal Eastern alluvial plain runs from Suvarnrekha River in the north to Rushikulya River in the south. The northern plateau is an undulating country having a general slope from north to south. Notable hill peaks in this region are Malyagiri, Mankarnacha and Meghasani. The southern hill region comprises of wide-open plateaus fringed by forested hills and has the highest places in the state. The Mahanadi basin graben lies between the northern plateau and southern plateau. It comprises the watershed areas of the principal rivers of the state. The western plateau constitutes flat lands with undulation and straight hill ranges. The aforesaid divisions have been made on the basis of NNE-SSW and E-W mega lineaments (Fig. 3.2).

2.5 Surface Manifestation

2.5.1 Attri

The thermal spring at Attri is emanating from a single spot in the midst of paddy field. A circular tank with iron railings now encloses it (Fig. 2.2). The diameter of the tank is 2.5 M and it has a 2.2 M water column in it. Vigorous gas activity accompanies with the water discharge. No spring deposits are noticed around the enclosed or along the channel through which excess hot water flows out. The hot water then discharged into other tanks beside the main tank for bathing purpose. The spring is sulphurous in nature and the gas fills the air with sulphur smell.

2.5.2 Tarabalo

The thermal springs at this place emerge at an elevated area covering about 8000 m^2. Thermal activity is confined to several spots (at least 25) in number and patches of warm ground through which hot water oozes out. It is being considered as the largest thermal field in Odisha with highest temperature. Gas bubbles are seen with water but less frequent than that of Attri. There are wild growths of screw pine (*Pandanus* sp.) locally known as "Kiabani" near the hot spring area. The springs are sulphurous in nature. The area is swampy and at places due to soft soil one starts slowly sinking into mud (Fig. 2.3). Some patches of mud slurry are common and animals sometimes fall victim of them.

2.5.3 Deuljhori

About twenty loci of thermal water manifestation are located here. It is considered as a place of sanctity and hence is known by the name Deuljhori. Each spring is

Fig. 2.2 **a**, **b**, **c** Photograph showing Attri Hot spring complex and the spring enclosed by masonry structure at Attri

Fig. 2.3 **a, b**, **c** Photograph showing masonry structures around thermal spring cluster and swampy marshy land due to oozing of hot water in the open field at Tarabalo

enclosed by a rectangular masonry structure (Fig. 2.4). Luxuriant growth of screwpine (*Pandanus* sp.) is observed around these localities. Steam and other gases come out of the edifice of the spring.

Fig. 2.4 **a, b, c, d** Photograph showing masonry structure around Deuljhori thermal springs

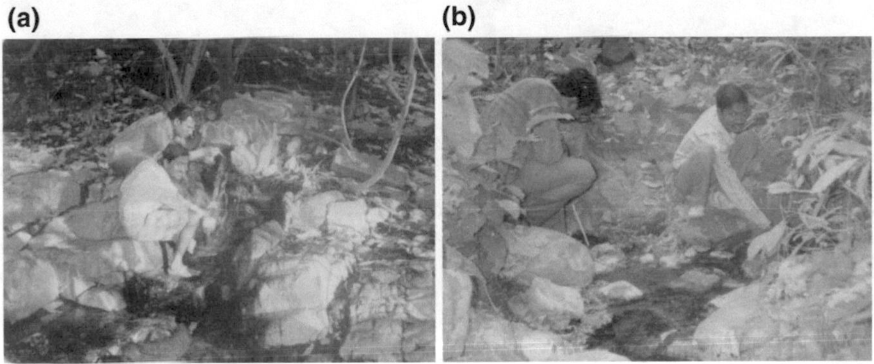

Fig. 2.5 Photograph showing oozing of hot water at **a** Magarmuhan and **b** Bankhol

2.5.4 Magarmuhan and Bankhol

At both the places the springs issue from a single vent (Fig. 2.5) from sheared quartzites belonging to Iron Ore Supergroup. The hot water discharge is continuous. These two springs smell no sulphur. The hot water flows into available colder drainage system of the area. These two springs have been located for the first time during the present investigation.

Fig. 2.6 Masonry structure around Taptapani thermal spring

2.5.5 *Taptapani*

The spring oozes from a single spot from inside the valley and is, as usual, encircled by a masonry structure (Fig. 2.6). The spring emanates along the Taptapani shear zone and the water flows into the valley. As is already mentioned it emits sulphurous smell.

2.5.6 *Boden*

The thermal spring at Boden ("Patal Ganga") oozes out from a single vent and is protected by a masonry structure (Fig. 2.7). The water is clean and odourless and no gas bubble is observed. The temperature is measured as the lowest among all the thermal springs of Odisha.

2.5.7 *Badaberena*

The hot spring near Badaberena in Angul district has been located for the first time during the present investigation. Hot water oozes out at three isolated spots covering an area of 1000 m^2. The spots are at ground level and on the western bank of N-S flowing Gauduni nala. The adjoining area is swampy due to the seepage of hot water. The water is clean but emits sulphurous smell. The spring is protected by a masonry structure (Fig. 2.8).

Fig. 2.7 **a, b** Masonry structure around Boden thermal spring

Fig. 2.8 Masonry structure around Badaberena thermal spring

2.6 Discussion

Present investigation incorporates locating and documenting the characters of thermal springs of Odisha. Till date eight thermal springs are known to occur in Odisha. The study and research during the period of work led the author to locate three (Badaberena, Magarmuhan and Bankhol) thermal springs and report for the first time. The thermal springs at Attri, Taptapani, Magarmuhan, Bankhol and Boden ooze out from single spot while that at Tarabalo, Deuljhori and Badaberena hot water come out from several vents. The ground surface at Tarabalo and Badaberena is swampy. The thermal springs at Attri, Tarabalo, Deuljhori, Taptapani, Badaberena exhale sulphur smell. This may be due to presence of sulphur bearing substance in the rocks through which the hot water percolated.

References

Acharya S (1966) The thermal springs of Orissa. The explorer, pp 29–35

Acharya S (2000) Some observations on parts of the Banded Iron-Formations of Eastern India. Pres. Address, 87th session, Ind Sc Cong Ass pp 1–34

Chapter 3
Geological Setting

Abstract This chapter deals with the geological and tectonic framework around the thermal springs. The springs are traced to occur along tectonic zones and in the absence of any active volcano, Odishan springs can only claim to be controlled by tectonic factor. These springs are either located in Gondwana graben or intersection of lineaments of Precambrian terrain. The rock assemblages around the thermal springs belong to the Iron Ore Supergroup, Easternghat Supergroup or Vindhyan Supergroup depending on their location as superimposed on the geology of the state. While the quartzites of Iron Ore Supegroup are traced around Magarmuhan and Bankhol region, khondalite, charnockite, augen-gneiss (granitic) are the litho-types encountered near the thermal springs at Attri, Tarabalo, Taptapani and Deuljhori. The rocks have suffered highest grade of regional metamorphism and are intricately folded and belong to Precambrian age. These Precambrian rocks are repository of radioactive minerals and possibly have led to increase radioactivity. Mineralogy of these rocks is discussed. The special features of the rocks of Gondwana Supegroup in so far as they control the different parameters of thermal springs are described and discussed. The rock types and their mineralogical composition are important as they play their roles during the percolation of meteoric water through them.

3.1 Introduction

Geological history of Odisha dates back to more than 3700 M.Y (Sarkar and Saha 1983) and the terrain is covered mostly by Precambrian rocks. Roughly about 25% are comprised of Post-Cambrian sediments including recent alluvium (GSI 1974).

The rocks of Precambrian age are represented by early granites and gneisses, charnockites, khondalites and para-metamorphites, and intrusives of ultrabasic-basic, doleritic and granitic nature. The Phanerozoic are represented by the rocks of Gondwana Supergroup and the Baripada beds belong to lower Tertiary (mio-pliocene) period. Laterites occur at different levels indicating peneplanation. The geological map of the state is shown in Fig. 3.1.

S. C. Mahala, *Geology, Chemistry and Genesis of Thermal Springs of Odisha, India*, SpringerBriefs in Earth Sciences, https://doi.org/10.1007/978-3-319-90002-5_3

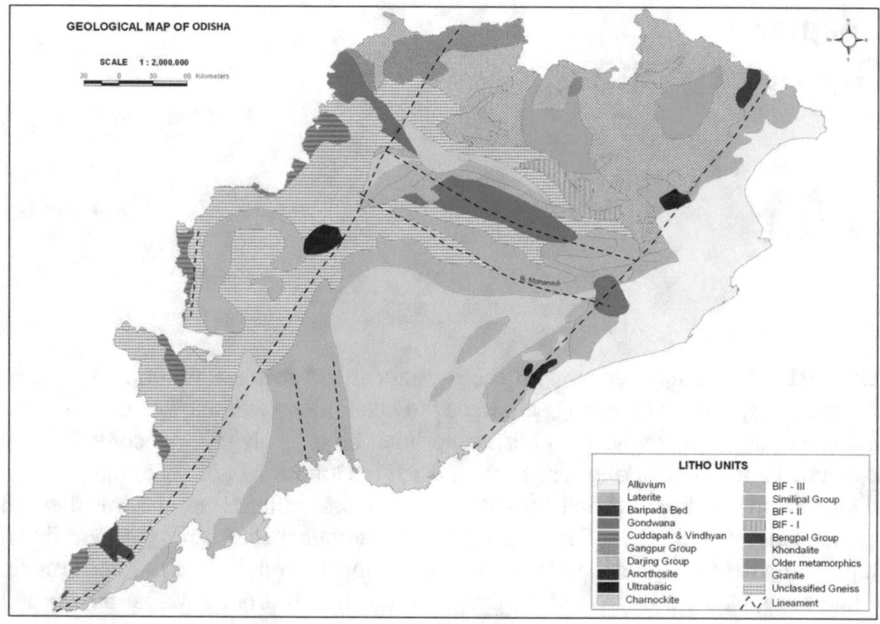

Fig. 3.1 Geological map of Odisha (GSI 1974) superposed by major lineaments to show possible tectonic control on thermal springs

Two important lineaments (NE-SW) and their secondary (size wise) counterparts (EW-forming Gondwana graben) seem to control the localization of the thermal springs in this region (Fig. 3.2).

According to Ellis and Mahon (1977) the geothermal fields in the Pacific basin particularly those in Japan, New Zealand, Indonesia and Chile are located in the zones of predominantly rhyolitic or andesitic volcanism. They further mentioned that wide spread hydrothermal activity in Iceland occurs in extensively fractured zone predominantly in basaltic rocks. In the Caenozoic tectonic regions, geothermal fields are located in different types of metamorphic and sedimentary rocks. Similar type of rock assemblages may produce different types of geothermal fields e.g. at Larderello, wells produce mainly dry steams whereas at Kizilidere, Turkey in a zone of similar rocks wells tap high temperature water. The known high temperature geothermal fields are often associated with Quaternary or recent volcanic activity, with faulting, graben formation, tilting, commonly nearby intrusion of rhyolitic rocks (Banwell 1970). Ellis and Mahon (1977) have mentioned a little about the Indian hot springs at Puga, west coast, and in the states of Bihar and Odisha.

In India, hot springs are observed in a wide variety of geological environments. The geothermal provinces of India are distributed along the following zones (1) Tertiary and Quaternary orogenic activity zones in the Himalayas, (2) Mesozoic block faulting zones (3) active seismicity and intense neo-tectonism zones (Shanker et al. 1991). The geothermal regimes are located in various litho-assemblages belonging to different geological period.

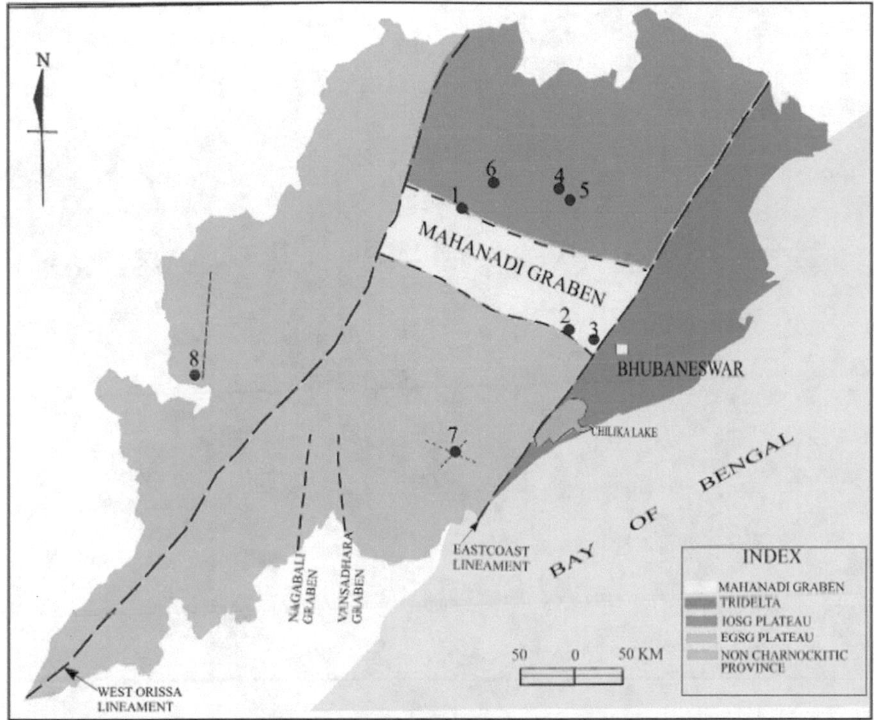

Fig. 3.2 Outline map of Odisha with mega lineaments showing location of thermal springs and major tectonic zones

Regional geological investigation around the thermal spring areas of Odisha leads to decipher the broad structural framework and to establish their relationship to the geothermal activity. Hence, they are mainly confined to crystalline schists or granite gneissic terrain of Easternghat Supergroup or Iron Ore Supergroup. Boden is only located within Vindhyan Supergroup. They do not seem to have any volcanic affinity as no recent/sub-recent volcanism is reported in and around Odisha.

3.2 Tectonic Framework

Certain geological conditions are necessary for the thermal springs to ooze out on the surface of the earth from deep-seated sources. As a matter of fact most of the hot springs are usually located in tectonic regime. They emerge more or less along a line, which indicates a mega-lineament, fault or fissure. These structural breaks serve as the avenues through which heated water from deep regions appear at the surface. The location of thermal springs are along distinct lineaments (Figs. 3.2 and 3.3) and

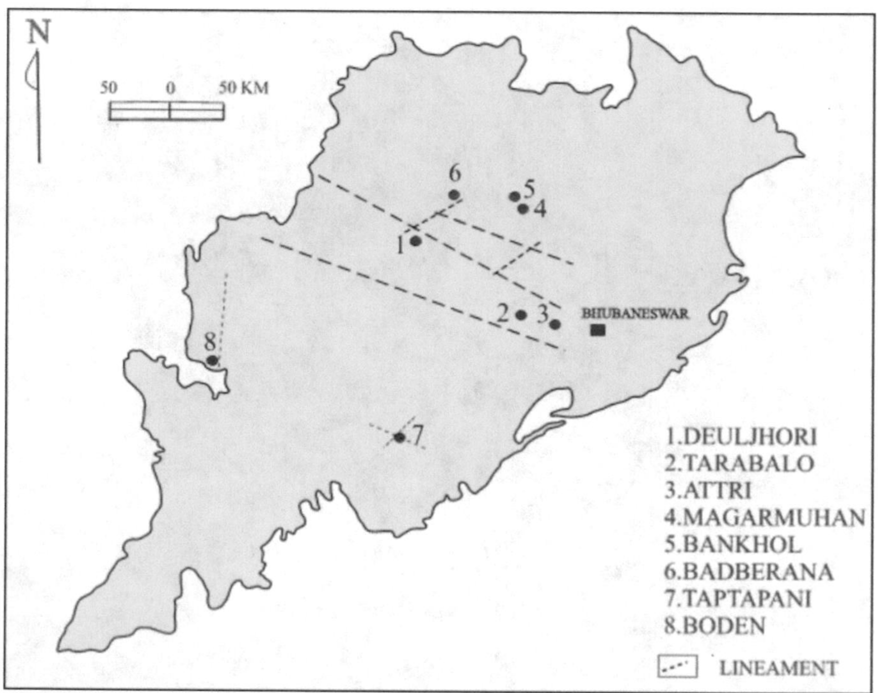

Fig. 3.3 Map showing lineaments related to thermal spring locations

these structural features are also remotely connected with thermal activities created possibly by deep seated intrusion related activity or any other heat producing factor along them.

During the course of recent work three new thermal spring areas have been traced at Magarmuhan, Bankhol and Barberena, which have not been reported earlier. The thermal springs at Magarmuhan and Bankhol are aligned in lines, which are more or less coinciding with the mega lineament traced from satellite imagery. The linear arrangement of the spots of thermal springs broadly matching with the trend of the lineaments suggests possible genetic relations of them. These thermal spring alignments are also discernible as the major lineaments in the satellite images (Fig. 3.4). The different major lineaments are identified as follows:

(i) Mahanadi lineament (Attri, Tarabalo, Deuljhori), southern fault of Gondwana basin.
(ii) Lineament in Easternghat granulite terrain (Taptapani), secondary lineament parallel to east coast lineament.
(iii) Lineament in Iron Ore Supergroup terrain (Magarmuhan and Bankhol), secondary lineament parallel to the northern fault of Gondwana basin.
(iv) Lineament along boundary of Chhattisgarh basin (Boden) and
(v) Lineament along Gondwana-Precambrian boundary (Badaberena), northern fault of Gondwana basin.

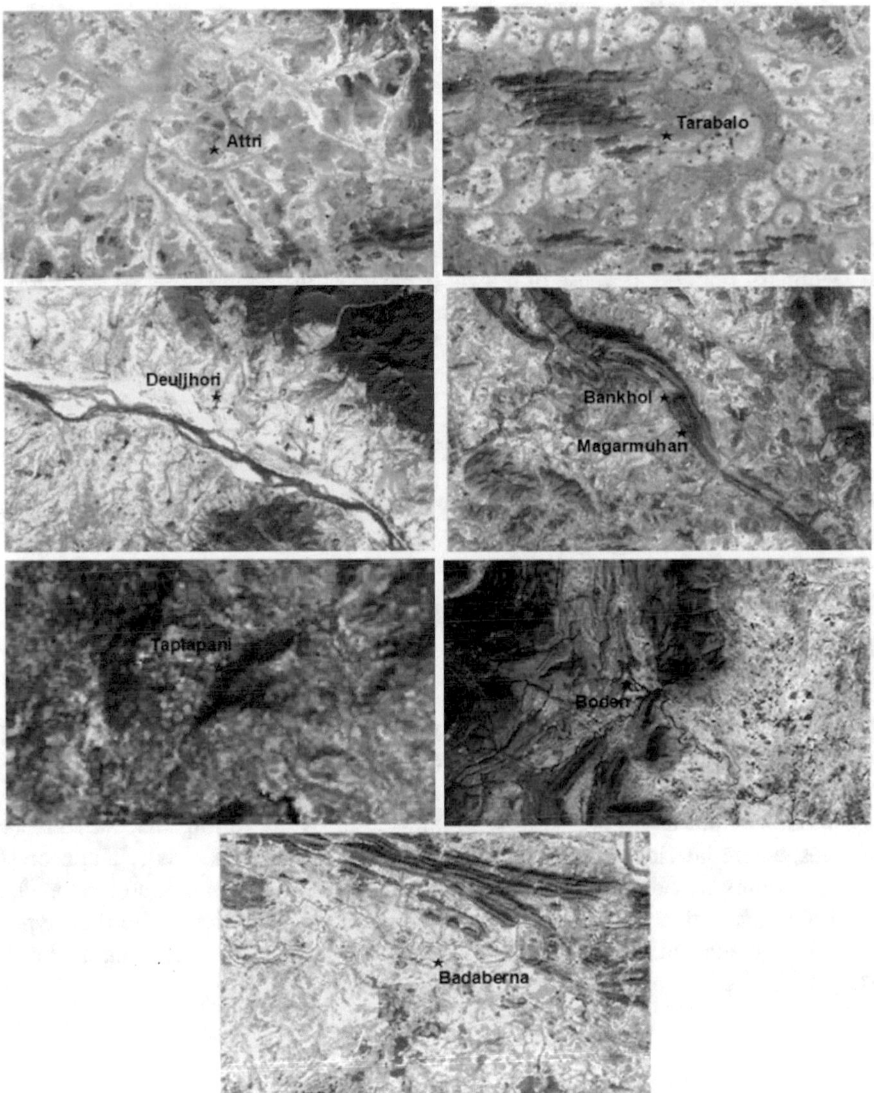

Fig. 3.4 Satellite image (FCC) of surrounding area with superimposed thermal spring locations

Satellite image data indicate the thermal springs of Attri, Tarabalo, Taptapani, Deuljhori and Badaberena fall as mentioned earlier on or near the intersection of two or more lineaments. The thermal springs at Attri and Tarabalo lie on the WNW-ESE trending major lineament parallel to the Mahanadi graben. Attri is located on the intersection of NNE-SSW trending Rana river lineament and WNW-ESE trending lineament (parallel to Mahanadi lineament). At Tarabalo the springs fall along NNE-SSW lineament at a small distance from another major NW-SE lineament. Detailed study on the satellite image indicates that the area around Attri exhibits a grayish tone on the FCC (False colour composite). The area shows intense land use for agriculture whereas the southern portion is lateritic plains and barren.

At Tarabalo, on the FCC, the area shows light grey tone, zigzag pattern indicating valley-fills with a rough texture. These valley fill areas without forest naturally intensely used. In the south and north of Tarabalo the area shows lighter tone of barren lands (lateritic uplands). On satellite image the area around thermal springs at Deuljhori, possibly because of oozing of water at different places shows a little different tone and texture from the surrounding area.

3.3 Litho-Assemblages

The lithologies of the rocks nearest to the thermal springs have been briefly described below.

3.3.1 Attri

The thermal spring at Attri lies in a low-lying area almost on the Rana river valley within a lateritic cover (Fig. 3.5). Rock exposures are lacking within a radius of 3 km around the spring where alluvium and diluvium are present. The uplands (with lateritic mass) are noted near Malisahi, Parasbasta, Attri, Kerotgadia, Saradhapur villages, where lateritic slabs are being quarried as building materials. The nearest rock exposures are observed 3–4 km south of the thermal spring around Bhatapada, Godisahi and further towards southwest and southeast directions. The rock types around Rautpada and Sovadei hill comprise mainly of khondalites and charnockites. They trend in east-west direction with sub-vertical banding.

3.3.1.1 Khondalite

Khondalite, is the predominant lithotype of Easternghat Supergroup. It is essentially composed of quartz, feldspar, garnet, sillimanite with or without graphite and it shows well marked schistose/gneissose fabric. The rock has suffered high-grade metamorphism up to granulite facies (Dash et al. 1987). Presence of dark brown colour garnets, elongated quartz and feldspars is noted in the weathered profile. Sillimanite needles are elongated along the foliation planes. Garnet porphyroblasts are corroded and exhibit sub-rounded shape.

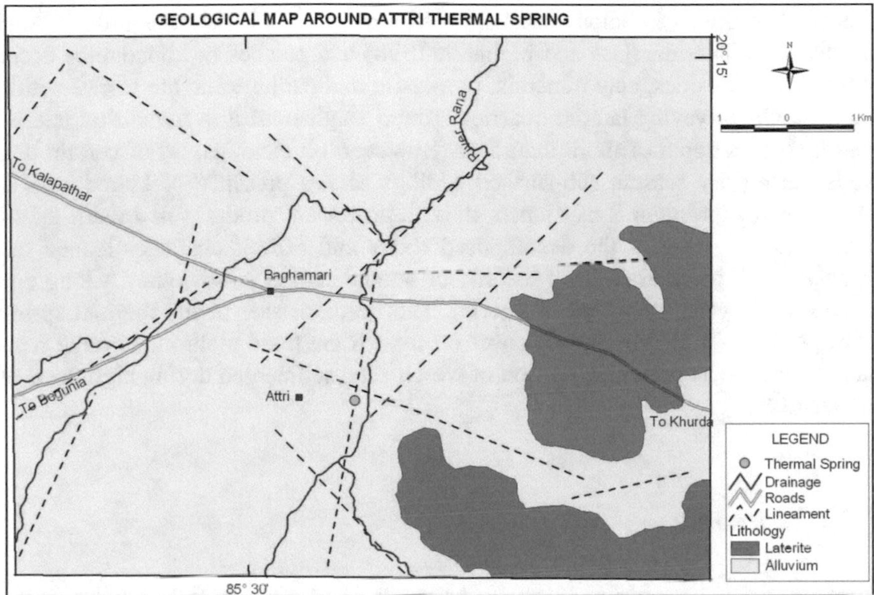

Fig. 3.5 Geological map around Attri thermal spring

3.3.1.2 Charnockite

Charnockite is the other important litho-unit of Easternghat Supergroup. It comprises of quartz, feldspar, garnet and ortho-pyroxene, and shows granulose texture. It is considered as both ortho-and para-metamorphites and suffered granulite facies of metamorphism. It contains almost fresh, grey to bluish grey colour quartz, white feldspar and pinkish brown garnet. Though they show granulitic texture but on weathered surfaces, a faint foliation is marked with identical attitude with that of khondalites.

3.3.1.3 Leptynite

Leptynite is also encountered in the area and consists of quartz, feldspar and garnet. It is leucocratic in hand specimen and shows granulose texture. Garnet of pinkish brown colour is well noted in this rock. Leptynite in the study area occur as intrusions within khondalites and charnockites in form of veins.

3.3.1.4 Laterites

Laterite is a secondary rock on the surface and is the weathered product of parent rocks lying below. It is porous and cavernous in appearance but hard in nature.

It is reddish brown in colour because of iron staining. The laterites around Attri are low-level laterites (Das and Mahallik 1998) and patches of khondalites occur within them. Besides, clay minerals, pyrolusite and psilomelane are traced within laterites. On surveying laterite quarries around Baghamari it is found that laterite continues to a depth of more than 5 m. However, on examination of certain dug wells during dry season, sub-surface geology shows presence of khondalites at depth about more than 8 m. Hence, these laterites are products of in situ lateritisation at low level of the decomposed rocks and exhibit character loaned out by their fresh parental rocks. The time of peneplanation is obviously a long one as appeared by the thickness of laterite. The northern side of the thermal spring shows a thick (8–10 M) alluvial cover on river Rana flood plain may be the remnant of river Rana plateau, a portion of which even submerged during high flood of Mahanadi.

3.3.2 *Tarabalo*

Tarabalo thermal springs exposed to the north of Mundipara hill and are on the pediment. The cluster of thermal springs is lying along the valley of Kaligiri nala. These thermal springs lie at a 2–3 m higher elevation than the surrounding ground carved out of valley fill.

The area around the thermal springs is swampy, muddy land where hot water oozes at different spots. The litho-assemblages comprise of charnockites and porphyroblastic granite gneisses belonging to Easternghat Supergroup (Fig. 3.6). The area adjoining to thermal springs is covered with soil/alluvium.

3.3.2.1 Charnockite

Charnockite is exposed in nearby area of Tarabalo thermal spring (Fig. 3.7 a, d) and comprises of quartz, feldspar, garnet and mafic minerals mostly pyroxene (Sahu 1974). It is hard, compact and marked with incipient joints/fractures and metamorphosed to granulite facies.

3.3.2.2 Porphyroblastic Granite Gneiss

Porphyroblastic granite gneiss is another important rock observed in and around the area (Fig. 3.7 b, c). The porphyroblasts are greyish white in colour, more or less lensoidal in shape and can be termed as augen gneiss. Their origin is thus clear.

Some porphyroblasts are elongated in habit and the longer axes are often more than 8 cm and the shorter axes measure 3–4 cm. The elongated porphyroblasts are identified as microclines. The host rock, designated, as palaeosome mainly comprises of pyroxene granulites and has a broad foliation strike east west with sub-vertical dip.

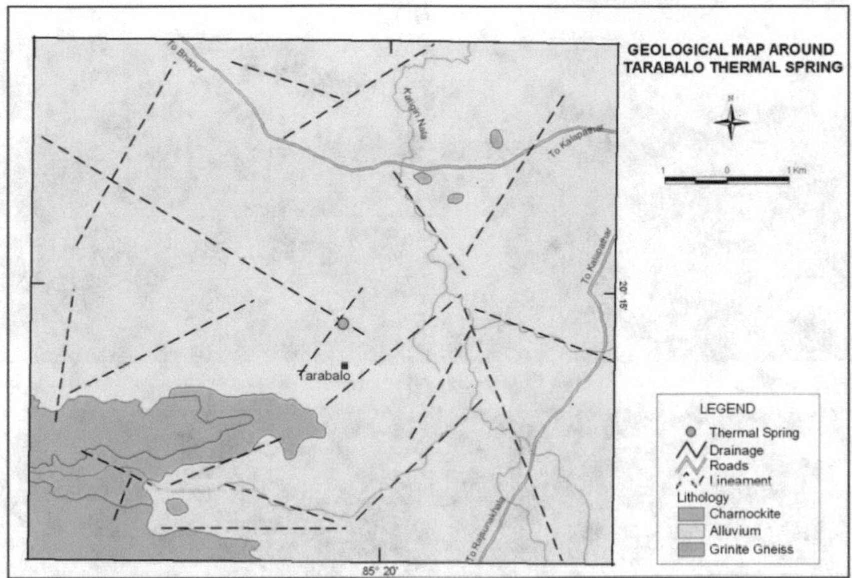

Fig. 3.6 Geological map around Tarabalo thermal spring

The feldspathisation is rampant and the rock has been converted to porphyroblastic augen gneiss. To the west of village Begunia and Golapokhari identical gneissic and granulitic rocks are exposed. The elongated hill located on the SSW direction of the thermal spring composed of khondalites, pyroxene granulites and leptynites. Khondalite in this area is well banded having east-west trending foliation with sub-vertical dip due north. The basic granulites and leptynites are not much weathered and exhibit faint foliations only on weathered surfaces.

3.3.3 Deuljhori

The thermal springs are exposed in an almost flat surface covering an area of about 3000 m^2. The litho assemblage of the region consists of two distinct groups. The first type comprises of Khondalites, garnetiferous quartzites and acid gneisses (porphyroblastic) of Easternghat Supergroup whereas the second type consists of sandstones and siltstones of Gondwana Super group (Fig. 3.8). The thermal springs are located almost at the contact zone between Pre-Cambrian and Gondwana formations. Easternghat Supergroup of rocks are exposed on the Panchadhar ridge trending NW-SE on the north-eastern side of the thermal springs at a distance of about 4–5 km.

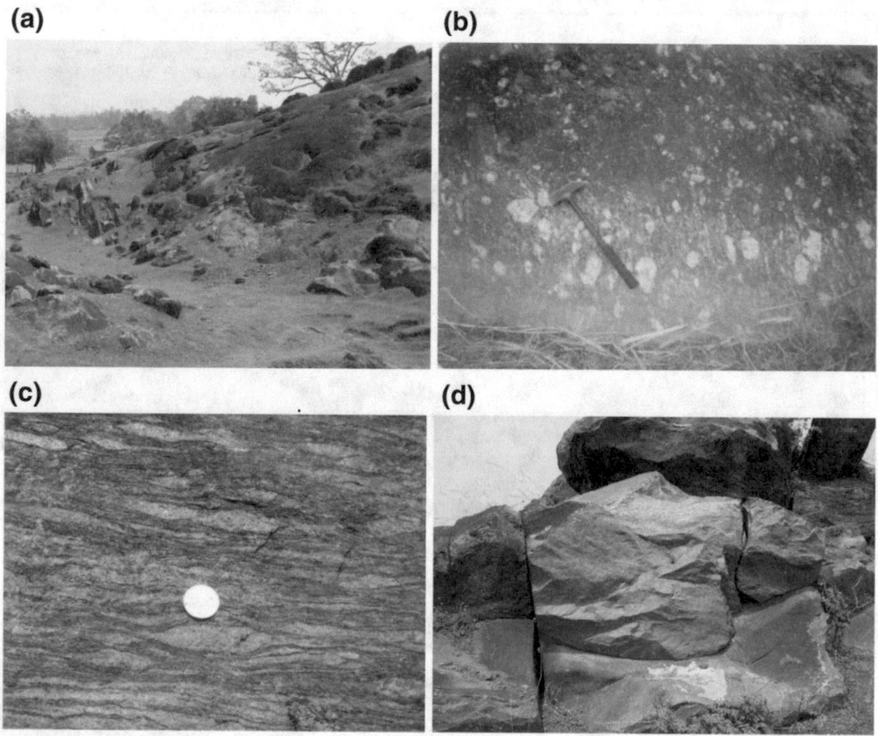

Fig. 3.7 **a** Photograph showing massive disjointed charnockite exposed in a small hillock near Tarabalo. **b** Photograph showing elongated and oriented phenocrysts of feldspars suggesting pre or syn magmatic orientation. Both coarse and medium grained phenocrysts are well noted near Tarabalo. **c** Photograph showing flattened augen structure containing lens shaped plagioclase feldspar in granite gneiss near Tarabalo. **d** Photograph showing fresh surface of charnockite exposure near Tarabalo

3.3.3.1 Acid Gneiss

The banded acid gneiss (porphyroblastic) is exposed on the north and eastern side of the thermal spring within a distance of 200 m (Fig. 3.9a). The porphyroblasts are commonly pink coloured microclines. A faint foliation is marked within the basic granulite. In the south, patches of basic granulites are exposed sporadically. The porphyroblastic gneisses are well exposed in Dandatapa nala. The porphyroblasts are mainly augen-shaped with the longer axes extend for more than 10 cm. Often these porphyroblasts are twinned and square-shaped. These porphyroblasts contain large number of inclusions of garnets. Quartz veins and pegmatites are emplaced along the foliation plane.

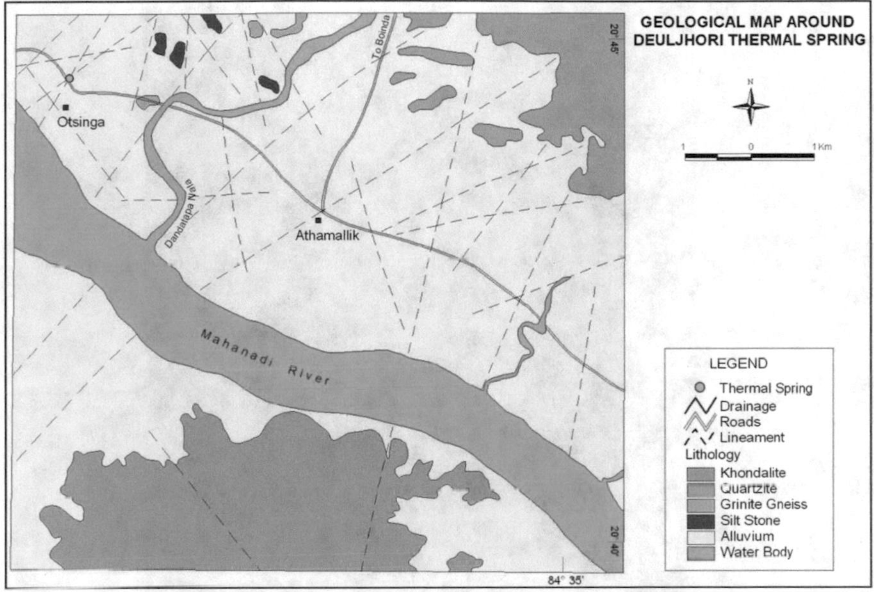

Fig. 3.8 Geological map around Deuljhori thermal spring

3.3.3.2 Sandstones and Siltstones (Gondwana Rocks)

On the riverbed of Dandatapa nala sedimentary rocks of Gondwana Super group are well exposed near Jiradghat (Fig. 3.9 b, c). The rock types consist of sandstone, siltstone and shale. Some boulders are traced in the area and composed of khondalite of variable size.

The sandstone is coarse grained in nature and variable in colours. They are marked with current bedding and ripple marks. The siltstone is the fine-grained variant between the underlying sandstone and the overlaying shales, which also exhibits impressions of current bedding. The shales are variegated but dominantly greenish in colour. They are splintery and break into long thin stick-like that resembles the greenish needle shales of Talchir group. At places the shales are also ferruginous (reddish brown in colour) and compact. The Gondwana sediments exhibit variable dip (12°–65°). Near the contact zone with Pre-cambrians the dip is very high about 60° (Fig. 3.9b) because of faulted contact evidences of which are as follows:

(i) Contact of Precambrian and Gondwana formations within a distance of 20–30 m.
(ii) Khondalite boulders within Gondwana sediments close to the fault contact plane indicating very close provenance when fault assisted the falling down of boulders into the basin.
(iii) Very high dip of Gondwana rocks (against usual dip 10°–15°) show drag effect due to faulting.
(iv) Straightness of river course is parallel to the contact.

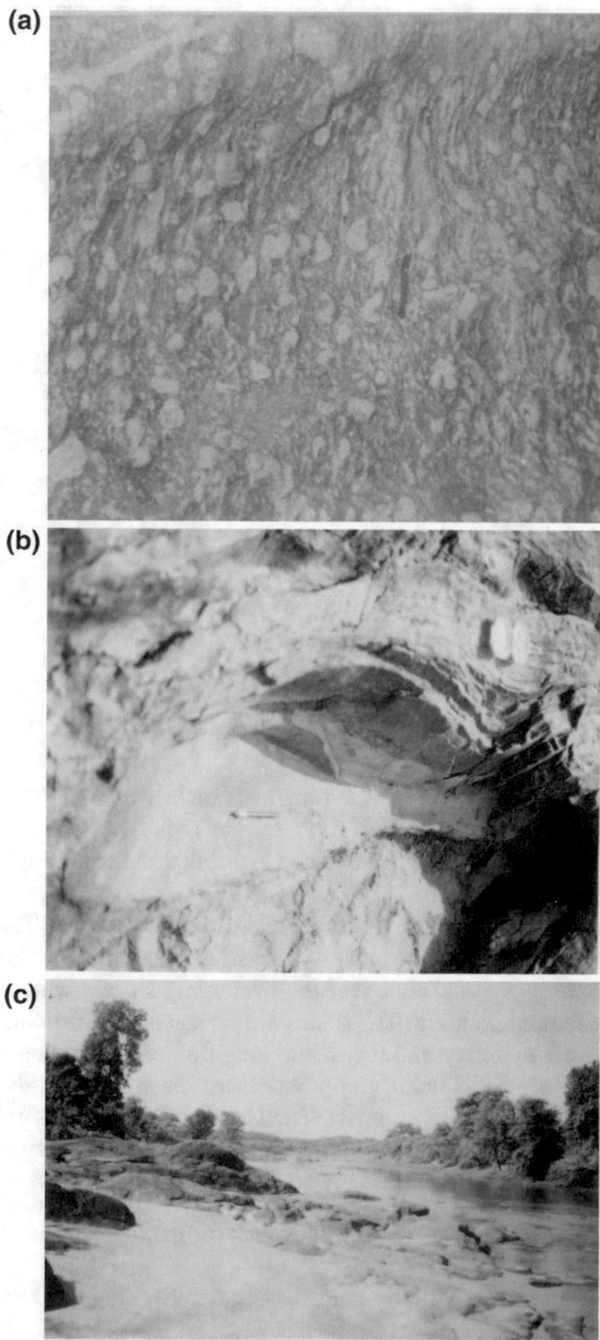

Fig. 3.9 **a** Photograph showing porphyroblastic granite gneiss exposed near Deuljhori thermal spring. **b** Photograph showing abnormal high dip of silt stones (Gondwana Group) near Deuljhori thermal spring. **c** Photograph showing contact of Gondwana rocks with Archaean rocks

The thermal spring also exists along the fault line having very high dip. This is specially so when it forms a long graben.

3.3.4 Taptapani

At Taptapani hot water oozes out from a single spot in the valley region lying on the eastern side of the state highway joining Berhampur and Rayagada. The Taptapani thermal spring is situated on the baseline of the contact zone between Taptapani-Mohana hill ranges and Digapahandi hill ranges. According to Swain and Padhi (1986) the thermal spring falls along a major NE-SW trending shear zone. The hill ranges comprise of rocks of Easternghat Supergroup. The litho units are khondalites, basic granulites, charnockites, quartzites, leptynites, granites and gneisses. Study indicates that the thermal spring lies in the vicinity of migmatised zone of Khondalite and Charnockite along the valley through which the shear zone passes. The charnockites crop out within a few meters of the spring and comprise of equigranular quartz, feldspar and pyroxene. Khondalite is exposed at higher altitudes than the thermal spring proper and composed of quartz, feldspar, sillimanite and garnet. Foliation and lineation features are well marked on the khondalites and charnockites.

A number of lineaments can be traced from the satellite image data of the area and a major shear zone is clearly traced in the field. The shearing is responsible for the crush breccias along the valley section including pseudo-tachylites. The hot water emanates along the major shear zone running NE-SW direction within the sheared charnockites (Fig. 3.10).

3.3.5 Magarmuhan and Bankhol

The thermal springs at Magarmuhan and Bankhol are situated in the western flank of a NW-SE trending ridge extending from Petraghati in south to Karpurhudi in north. The elongated ridge constitutes the northern limb of a mega fold. The ridge predominately consists of quartzites and quartz-schists belong to Iron Ore Supergroup. Being resistant to weathering, the quartzites stand out as ridge against the valley, which is flat and covered with soil and alluvium (Fig. 3.11). The main rock types in the area are quartzite, quartz-mica schist, quartz-kyanite schist and sillimanite quartzite. There are patches of mica schist within the quartzites with prolific garnet and staurolite crystals. The quartzites and quartz schists are highly jointed and the bedding joint is more prominent. The rocks, at places, show effect of shearing as indicated by presence of sheared quartzite and crush conglomerates. On regional scale the schistosity trends northwest-southeast direction with sub-vertical dip towards southwest. These meta-sedimentary rocks extend from Kamakshyanagar to Palalahara region. These rock assemblages belong to the oldest Iron formation (BIF-I; Acharya 2000). At

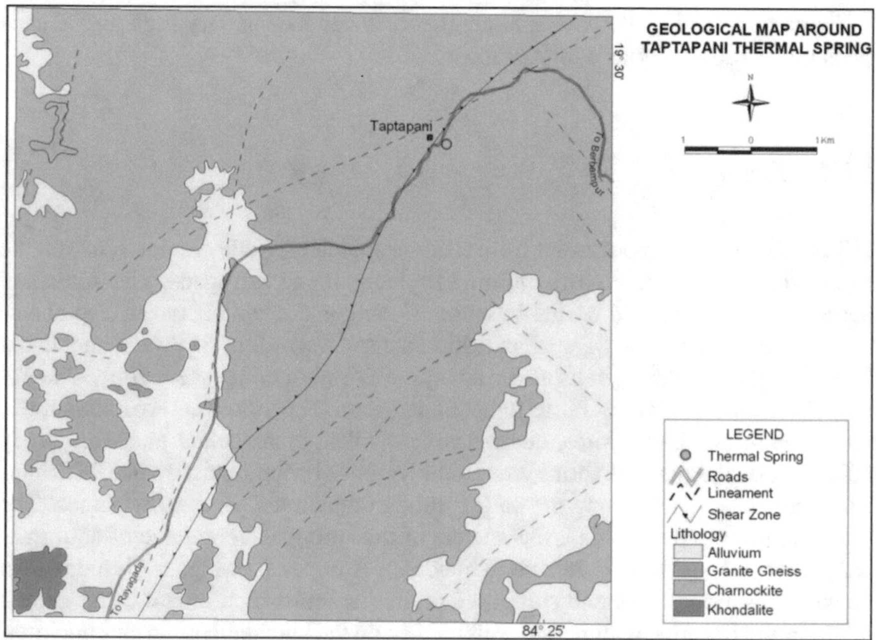

Fig. 3.10 Geological map around Taptapani thermal spring

Magarmuhan hot water emanates from jointed and fractured quartzites and quartz schists, at Bankhol from crush conglomerates.

3.3.6 Badaberena

The thermal spring at Badaberena oozes out on the bank of rivulet Gauduni, a tributary of Tikra River (tributary of R. Brahmani) (Fig. 3.13). The spring is located on the northern periphery of well-known Mahanadi valley graben (Pre-Cambrian–Gondwana boundary). Satellite image interpretation of the area enables to infer that the spring is situated at the intersection of two lineaments trending NW-SE and NE-SW directions (Fig. 3.13).

Highly sheared and brecciated quartzites with sheared quartz veins belonging to Tikra formation of Precambrian age (Mahalik 1994) are observed in the spring locality (Fig. 3.12). The quartzites are pinkish white to white in colour, tough and compact. Sandstones, siltstones and shales are located south of the thermal spring.

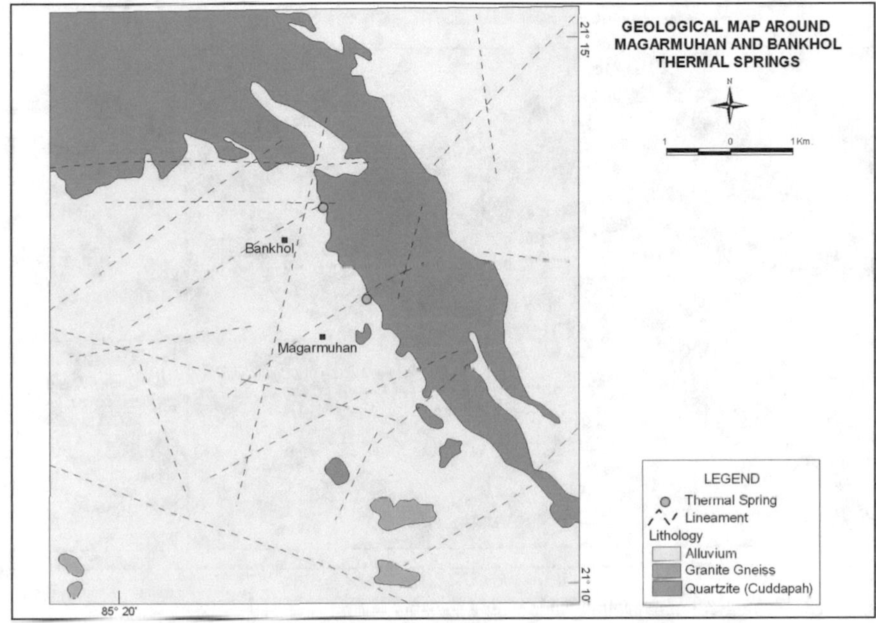

Fig. 3.11 Geological map around Bankhol & Magarmuhan thermal springs

Fig. 3.12 Photograph showing exposure of quartzites in Badaberena thermal spring area

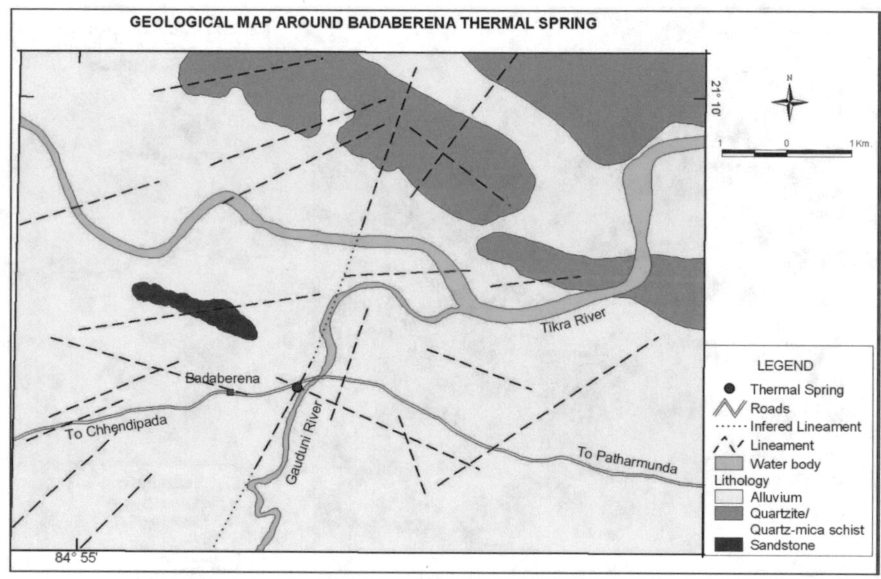

Fig. 3.13 Geological map around Badaberena thermal spring

3.3.7 *Boden*

At Boden, the thermal spring is emanating from a single spot on the south-eastern slope of Katpur hill (Acharya 1966). Local people name the spring as "Patal Ganga". Quartzite of Vindhyan Supergroup of Proterozoic age forms the adjoining hills. They are fully metamorphosed and the banding has resulted out of metamorphic differentiation (Stilwell 1922). The rocks of Easternghat Supergroup are exposed about 5 km away in NE direction from the thermal spring location. The thermal spring is located along a lineament trending N-S that is also incorporated from satellite image of the area (Fig. 3.14).

3.4 Discussion

Recent investigations on geological set-up in and around the thermal springs of Odisha have documented the following aspects on structure and tectonics, litho-stratigraphy and igneous activity.

The thermal springs are aligned more or less along the mega lineaments.

(i) They are mostly structurally controlled phenomena.
(ii) The thermal springs are confined to the rocks of Iron Ore Supergroup, Easternghat Supergroup and Vindhyan Supergroup rocks of Precambrian age.
(iii) No recent tectonic activity (neo-tectonism) in these areas is reported.

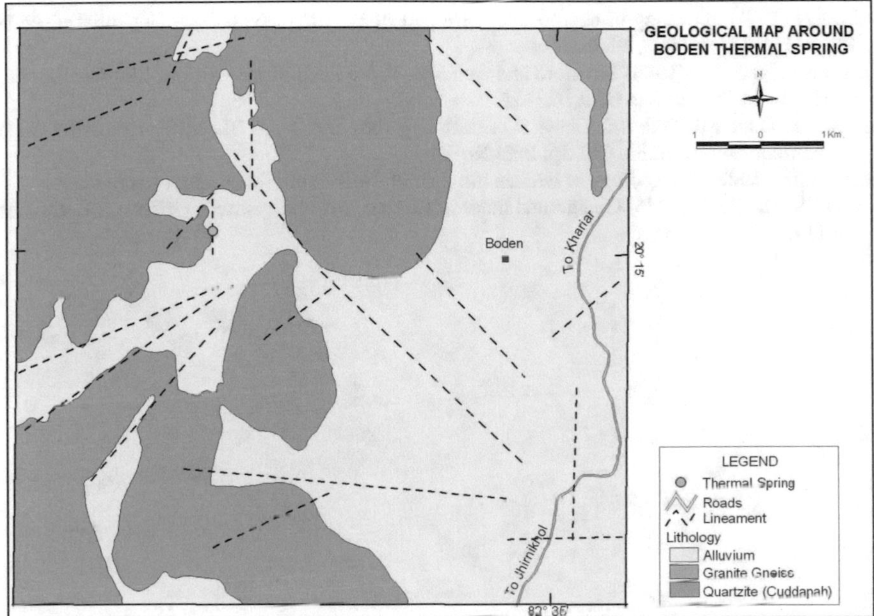

Fig. 3.14 Geological map around Boden thermal spring

(iv) There is also no information regarding recent/sub-recent volcanic activity associated with thermal springs.

(v) As mentioned earlier these water from depth move through the Precambrian rocks partly dissolved out their mineralogical constituents to the new system of association with thermal springs that resulted in enrichment of radioactive minerals.

References

Acharya S (1966) The thermal springs of Orissa. The Explorer. pp 29–35

Acharya S (2000) Some observations on parts of the Banded Iron-Formations of Eastern India. Pres. Address, 87th session, Ind Sc Cong Ass, pp 1–34

Das JN, Mahallik NK (1998) Geochemical evaluation of the laterites of Orissa. SGAT Bull 1(1):49–56

Dash B, Sahu KN, Bowes DR (1987) Geochemistry and original nature of Precambrian Khondalites in the Easternghats, Orissa, India. Trans Royal Soc Edinburgh Earth Sci 78:115–127

Banwell CJ (1970) Geothermics 1(special issue 2):32

Ellis AJ, Mahon WAJ (1977) Chemistry and geothermal systems. Academic press, New York

G.S.I. (1974) Geology and mineral resources of the states of India, Part-III, Orissa. Misc Publ No-30, pp 6–7

Mahalik NK (1994) Geology of the contact between the Easternghats belt and North Orissa craton India. J Geol Soc India 44(1):41–51

Sahu KN (1974) Petrology of charnockites of a part of Easternghats, Orissa. Unpublished Ph.D Thesis, Utkal University, Bhubaneswar

Sarkar SN, Saha AK (1983) Structure and tectonics of the Singhbhum—Orissa Iron-Ore Craton, Eastern India. Recent Res Geol 10:1–25

Shanker R, Guha SK, Seth SK, Ghosh A, Ghosh S, Nandy DR, Jangi BL, Muthuraman K (1991) Geothermal Atlas of India. GSI Spl Publ No-19

Stillwell FL (1922) The geology of Broken hill district, New South wales. Geol Sur Mem 8

Swain PK, Padhi RN (1986) Geothermal fields of Ganjam and Puri districts, Orissa. GSI, Records vol 114, Pt 3, pp 41–46

Chapter 4
Hydrological Characters

Abstract This chapter records the collection of field hydrological data during field study with the help of portable instruments. The different physical parameters of the thermal springs such as temperature, pH, colour, odour and rate of discharge of the thermal fluid are noted. The temperature of surface water of the thermal springs ranges from 32 (Boden) to 67 °C (Tarabalo). The pH value of the thermal waters shows a neutral to mildly alkaline (pH 7.4–7.8) character. All the thermal springs discharge colourless and clear hot water. Sulphur odour is discernible particularly at Attri, Tarabalo, Deuljhori, Taptapani and Badaberena. The volume of water released remains more or less constant for all springs and the rates of discharge recorded at different locations are different and is found to be the maximum at Magarmuhan.

4.1 Introduction

Thermal springs are characterized by water of variable temperature. This chapter deals with studies on the hydrological character of the thermal springs. Following the standard work of Day and Allen (1924), Day (1939), Ellis and Mahon (1977), Gupta et al. (1976), Vaselli et al. (2002) and Valentino and Stanzino (2003) it is accepted that the constancy on the pattern of temperature and discharge over a considerable time has a definite bearing on the source of the thermal water. Data on physical and hydrological characteristics of thermal springs such as temperature, pH, colour, odour and the rate of discharge are recorded during the field survey (Fig. 4.1) and given in the Table 4.1.

4.2 Temperature

Temperature is the most important character based on which the thermal springs are evaluated for their uses. The surface temperatures of water from different thermal springs are measured in the field itself with the help of laboratory thermometer. The

S. C. Mahala, *Geology, Chemistry and Genesis of Thermal Springs of Odisha, India*, SpringerBriefs in Earth Sciences, https://doi.org/10.1007/978-3-319-90002-5_4

Fig. 4.1 Photographs showing recording of physical characters of thermal springs

temperature of the spring water varies from 32 to 67 °C (Table 4.1), Boden being the lowest temperature (32 °C) and Tarabalo the highest (67 °C). The temperatures recorded at other springs are Attri 58 °C, Deuljhori 56 °C, Taptapani 42 °C, Magarmuhan 38 °C, Bankhol 44 °C, and Badaberena 50 °C. The temperature values are shown in a bar diagram (Fig. 4.2) graphically.

The hot springs along Mahanadi valley graben (Attri, Tarabalo, Deuljhori) indicate higher temperature than the other springs. It is interesting to note that during the three years of field investigations, the hot springs recorded a fairly uniform temperature throughout. However, variation of 2 to 3 °C in the water temperature is not uncommon. The thermal spring water does not show the influence of diurnal variation of temperature. Steam is observed in high temperature springs like Attri, Tarabalo, Deuljhori and Taptapani (Fig. 4.3). The author is conscious that a 2-3 years period may not be enough for drawing conclusions on temperature variation. But results of previous workers indicate that the variation is not more than 2-3 °C.

Table 4.1 Hydrological characters of thermal springs of Odisha

Name of spring	pH	Temperature in °C	Type of discharge	Odour
Attri	7.7	58	Hot water with steam	Sulphurous
Tarabalo	7.8	67	Hot water with steam	Feebly sulphurous
Deuljhori	7.4	56	Hot water with steam	Sulphurous
Taptapani	7.5	42	Hot water with steam	Sulphurous
Magarmuhan	7.7	38	Hot water	Odourless
Bankhol	7.5	44	Hot water	Odourless
Badaberena	7.8	50	Hot water with steam	Feebly sulphurous
Boden	7.4	32	Hot water	Odourless

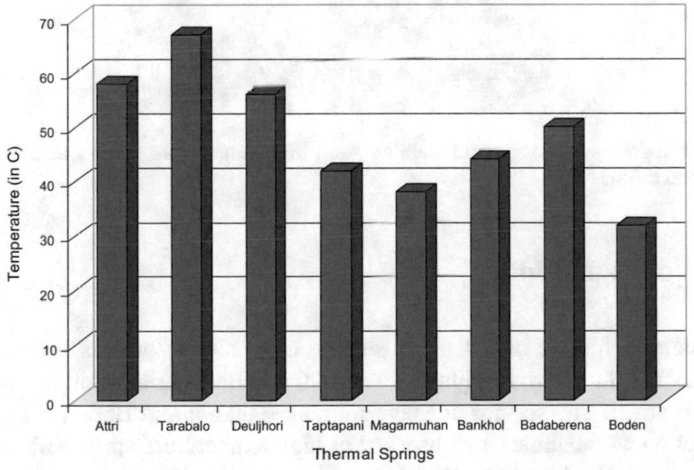

Fig. 4.2 Bar diagram showing temperature of different thermal springs of Odisha

4.3 pH (Acidity and Alkalinity)

It is defined as the negative logarithm of hydrogen ion concentration in water. Normal pH value of drinking water is 7 (considered as neutral); pH value below 7 and above 7 is considered as acidic and alkaline respectively. The pH of all the spring water was measured in the field itself with the help of portable pH meter. The pH values of different spring water vary from 7.4 to 7.8 (Table 4.1) that indicate a neutral to mild alkaline nature of the thermal water. According to Ellis and Mahon (1977) the water collected at the surface are of a higher pH than the deep high temperature water.

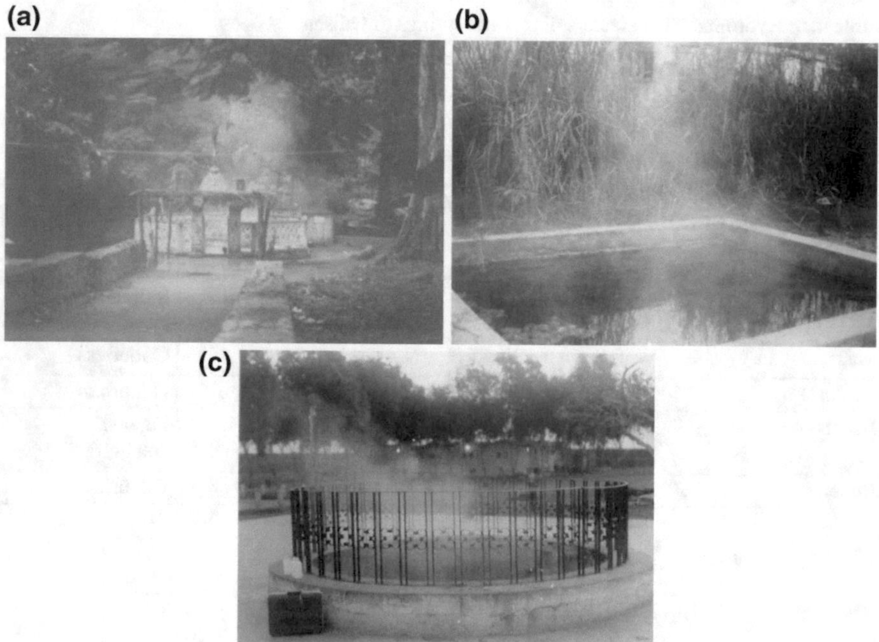

Fig. 4.3 Photographs showing steam coming from high temperature springs at **a** Taptapani, **b** Deuljhori and **c** Attri

4.4 Colour and Odour

All the thermal springs of Odisha discharge clear and colourless water. Thermal springs at Attri, Tarabalo, Deuljhori, Taptapani and Badaberena emit hot water with sulphurous smell. The springs at Magarmuhan, Bankhol and Boden are odourless. Steams get on with bubbles are observed in high temperature springs at Attri, Tarabalo, Taptapani, Deuljhori and Badaberena and cause micro-rings of waves on the surface of water (Fig. 4.4). The upsurging of gas bubbles is rhythmic at Attri and a great number of bubbles come every 40th or 50th second. Acharya (1966) reported presence of blue green algae in Attri thermal spring. The algae according to him belong to genus *Protococcoales* and species *thermopiles*. During present investigation white thin fibrous mass and pale yellowish encrustations are noted on the walls of masonry structures at Taptapani and Attri respectively. At Deuljhori and Tarabalo white to light yellow encrustation is also marked on the masonry structures and on the ground surface respectively (Fig. 4.5). There are luxuriant growth of screw pine adjoining to thermal spring areas at Deuljhori and Tarabalo. It is interesting to note that some tadpole like worms are swimming even in high temperature water as observed at Tarabalo.

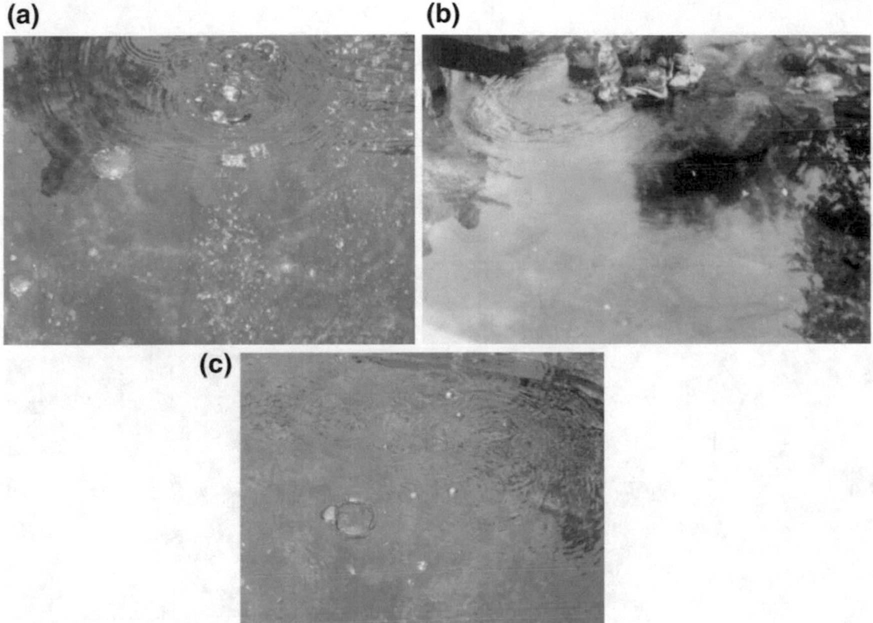

Fig. 4.4 Gas bubbles coming out from thermal springs at **a** Attri, **b** Taptapani, **c** Deuljhori springs

4.5 Rate of Discharge

Water discharges from the thermal springs are variable. The volume of water flow (liter) per unit time (minute) is known as rate of discharge. The rate of discharge of thermal spring water was measured by pouring the water into a graduated container of known capacity and noting the time duration required filling it up. Rate of discharge of hot water from the thermal springs was measured for individual thermal springs. The spring water at Taptapani, Attri and Magarmuhan indicate higher rate of discharge of 33, 62, and 84 lpm respectively. At Tarabalo and Deuljhori it was not possible to measure the rate of discharge as the hot water emanating at a number of spots and flow of water is very less. The rate of discharge at Boden and Badaberena could not be measured, as the orifice is not visible. The overall rate of discharge from the springs, apart from seasonal variations, remains nearly constant.

4.6 Discussion

All the physical characters of the eight springs have been recorded during the study. The highest temperature is recorded as 67 °C at Tarabalo followed by Attri and Deuljhori being 58 and 56 °C respectively. These springs, on the basis of temperature,

Fig. 4.5 **a** White encrustation on the ground surface near Tarabalo thermal spring. The encrustation may be cryptocrystalline silica due to dehydration of mineralized thermal water. **b** White encrustation on a rock surface inside Taptapani thermal spring. **c** Yellowish white encrustation on the surface of parapet wall of the masonry structure of Tarabalo thermal spring. **d** White encrustation on the surface of parapet wall of the masonry structure of Deuljhori thermal spring

are categorized as low enthalpy (Gropper and Schochet 1996) as thermal springs having temperature below 160 °C are considered as low enthalpy type. The pH value of water indicates mild alkaline to neutral nature, hence is suitable for domestic purpose. The water of the springs is colourless. Some springs imparting typical odour of hydrogen sulphide, indicate that the hot water absorb sulphur from the rocks during percolation. The rate of discharge has been recorded, the highest being at Magarmuhan (84 lpm). The flow of hot water depends on the amount of recharge and the nature (size) of the orifice; in fact this is due to the structural disposition (fault/joint controlled) of the area. Continuous flow of water from the springs reveals their perennial nature.

References

Acharya S (1966) The thermal springs of Orissa. The Explorer. pp 29–35

Day AL (1939) Presidential address to Geol Soc America on hot spring problem. Bull Geol Soc Am vol 50

Day AL, Allen ET (1924) Symposium on temperature of hot springs and the sources of their heat and water supply. J Geol 32

Ellis AJ, Mahon WAJ (1977) Chemistry and geothermal systems. Academic press, New York

Gropper J, Schochet DN (1996) Performance of geothermal combined steam/Binary cycle power plants with moderate and high temperature resources. GSI spl. Publ. No-45, pp 137–144

Gupta ML, Singh SB, Rao GV (1976) Geochemistry of thermal waters from various geothermal provinces of India. Int Asso of Hydrological sciences. Publ. 119, pp 47–58

Valentino GM, Stanzino D (2003) Source processes of the thermal waters from the Phlegraean fields (Naples, Italy) by means of the study of selected minor and trace elements distribution. Chem Geol 194. pp 245–274

Vaselli O, Minissale A, Tassi F, Magro G, Seghedi I, Ioane D, Szakacs A (2002) A geochemical traverse across the Eastern Carpathians(Romania): constraints on the origin and evolution of the mineral water and gas discharges. Chem Geol 182. pp 637–654

Chapter 5
Geochemistry of Thermal Water

Abstract In this chapter data derived from chemical analysis of thermal water are recorded and interpreted. The chemical composition of water is naturally different from one another. The predominant cations are sodium, potassium, magnesium and calcium whereas the anions include chloride, sulphate, carbonate and bicarbonate. On the basis of cation and anion percentage, the waters of thermal springs are categorized as NaCl, $NaHCO_3$ and $CaHCO_3$ types. The thermal spring water of Attri, Tarabalo, Deuljhori and Taptapani are of NaCl type while that of Magarmuhan, Bankhol and Badaberena are $NaHCO_3$ type and that of Boden is $CaHCO_3$ type. The analysis data are plotted on Piper trilinear diagram, Stiff diagram Wilcox diagram and U.S salinity diagram in order to determine the quality of the water. The quality of water of the thermal springs has led to infer its suitability for the use in drinking and agricultural purposes. The total dissolved solids (TDS) value is low for the thermal springs of Odisha, which indicate a modest amount of mineral content. Nevertheless, presence of radioactive elements imparts a therapeutic character. Geothermometry with respect to Na/K and Na–K–Ca geothermometer are carried out to determine the reservoir temperature of the thermal springs. The above process led to calculate the highest base temperature at Tarabalo thermal spring area.

5.1 Introduction

Geochemical investigations of thermal spring water and associated gas phases can provide information regarding the nature of geothermal systems (Ellis and Mahon 1977). Geochemical information regarding composition, temperature and pressure of the fluids within the geothermal system are generally derived from the chemical analyses of liquid phase (water) reaching the surface in form of springs, mud pools, fumaroles etc. Importance of chemical composition of thermal water, as an aid for geothermal exploration, has been well established (White 1970; Truesdell 1976; Gupta et al. 1976; Ellis and Mahon 1977; D' Amore and Truesdell 1979; Saxena and Gupta 1982; Saxena 1983, Singh et al. 2004).

S. C. Mahala, *Geology, Chemistry and Genesis of Thermal Springs of Odisha, India*, SpringerBriefs in Earth Sciences, https://doi.org/10.1007/978-3-319-90002-5_5

Chemistry of geothermal water provides information regarding the extent of mixing of deep thermal water with shallow cold water; reservoir temperature and qualitative information regarding the rock types and idea on rock-water inter action. Chemical characters of thermal spring water often provide clues to their ultimate origin. Based on the knowledge of chemistry of geothermal system, an area can be selected or rejected to carry out further exploration activity for the development of geothermal reservoirs.

Studies on different aspects of chemistry of Indian thermal springs have been carried out from quite some time (Chaterjee 1969; Deb and Mukherjee 1969). However, systematic studies on chemistry of thermal water have commenced recently (Gupta and Sukhija 1974; Gupta et al. 1975a, b; Gupta and Saxena 1979; Saxena and Gupta 1982, 1985, 1986; Saxena and D' Amore 1984; Singh and Bandopadhyay 1995; Bandopadhyay and Nag 1989). Study on chemistry of some thermal springs of Odisha has been carried out but of limited extent (Swain and Padhi 1986; Bhargav et al. 2001).

The present investigation deals with geochemical study and interpretation of the thermal springs of Odisha. Study includes analyses of the major elements such as Sodium (Na), Potassium (K), Calcium (Ca), Magnesium (Mg), Sulphate (SO_4), Chloride (Cl), Carbonate (CO_3) and Bicarbonate (HCO_3). It also includes electrical conductivity as a measure of total amount of salts/total dissolved solids (TDS).

Fig. 5.1 Photographs showing collection of water samples from different thermal springs

5.2 Methodology

Water samples from all the eight thermal springs were collected in one-litre capacity white polypropylene screw capped bottles (Fig. 5.1). The bottles were cleaned and rinsed two to three times with the spring water first, and then filled with respective water sample. The bottle caps were fixed tightly and sealed properly with wax and were taken to the laboratory for analysis. pH and electrical conductivity (EC) were determined on the spot with portable water analyzer kit (Fig. 5.2).

Cations like sodium and potassium were determined by flame photometer (Systronics-digital). Calcium and magnesium were determined by conventional EDTA titration methods. Anions like carbonate and bi-carbonate were calculated nanographically by acidmetry method (APHA 1995). Chloride was determined by Mohr titration method and sulphate by turbidmetric method using a sprectrophotometer having light path 2.5 cm.

Fig. 5.2 Photograph showing measurement of different physical parameters of thermal water

Table 5.1 Chemical composition of thermal water (Concentrations in mg/l)

Spring Name	EC	TDS	Na	K	Ca	Mg	Cl	SO_4	CO_3	HCO_3
Attri	987	560	174	4	22	2	254	60	4	14
Tarabalo	762	426	126	3.5	6	2	184	47	6	24
Deuljhori	1055	642	184	6	35	2	248	64	8	58
Taptapani	345	214	38	2	22	4	60	20	14	42
Magarmuhan	382	238	35	6	22	4	46	2	16	82
Bankhol	348	216	32	3	20	4	48	2	12	84
Badaberena	406	232	29	2	28	10	44	18	10	78
Boden	424	246	15	3	32	18	28	55	12	72

5.3 Results of Analysis

The results of chemical analysis of major cations and anions of thermal spring water from all the localities are given in Table (5.1). The results indicate that the water quality from all the eight thermal spring areas is not identical to one another chemically. Different types of water have been categorized on the basis of comparative study on the concentration of different chemical constituents present in the water. The concentration of cations and anions of the spring waters are represented graphically by bar diagrams (Figs. 5.3, 5.4, 5.5, 5.6, 5.7, 5.8, 5.9, 5.10, 5.11 and 5.12) and Pie diagrams (Figs. 5.13, 5.14, 5.15, 5.16, 5.17, 5.18, 5.19 and 5.20). Distribution of major cations and anions in the spring water can be summarized as follows.

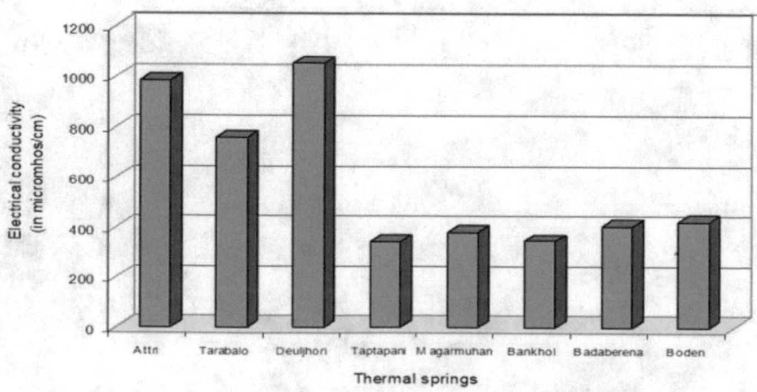

Fig. 5.3 Bar diagram showing variation of EC of different thermal water

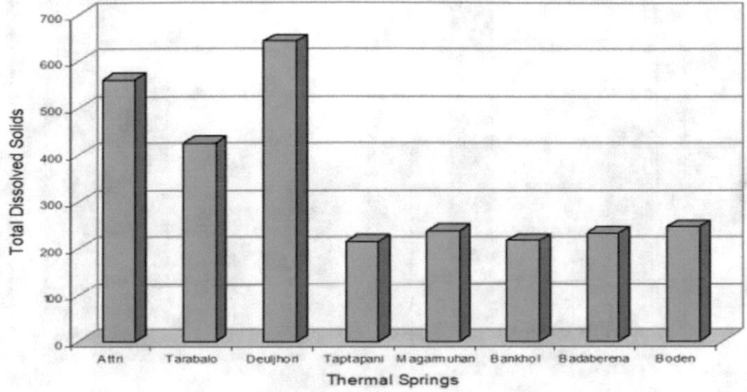

Fig. 5.4 Bar diagram showing variation of TDS of different thermal water

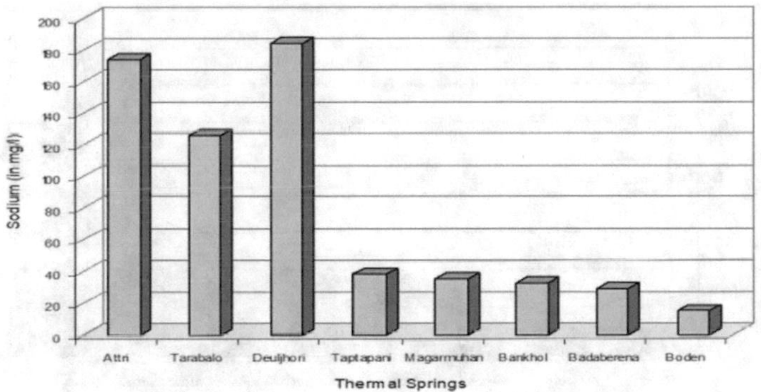

Fig. 5.5 Bar diagram showing variation of Na Conc. of different thermal water

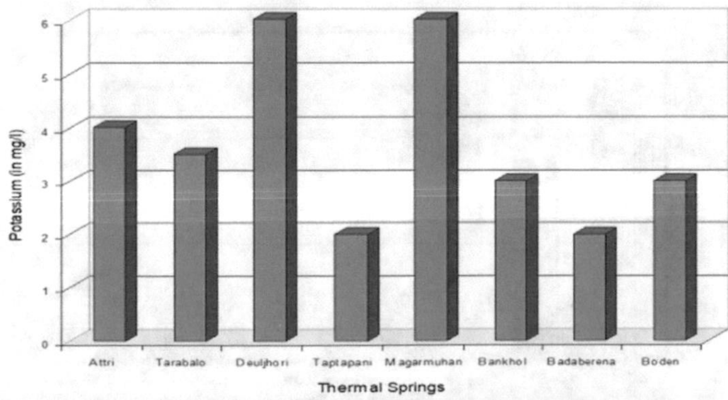

Fig. 5.6 Bar diagram showing variation of K Conc. of different thermal water

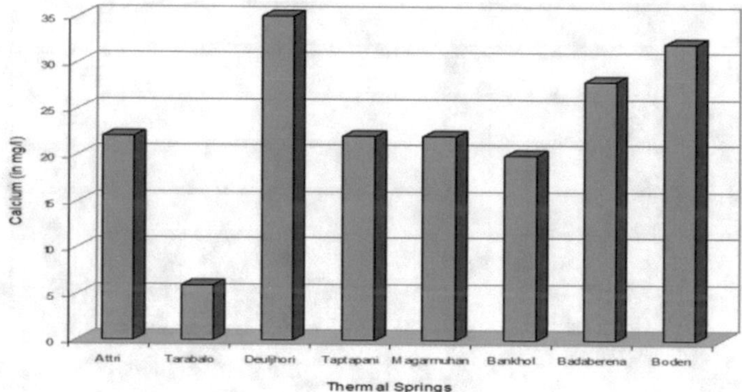

Fig. 5.7 Bar diagram showing variation of Ca Conc. of different thermal water

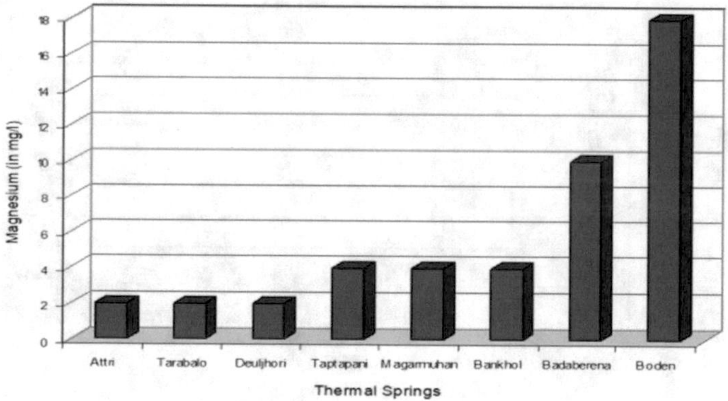

Fig. 5.8 Bar diagram showing variation of Mg Conc. of different thermal water

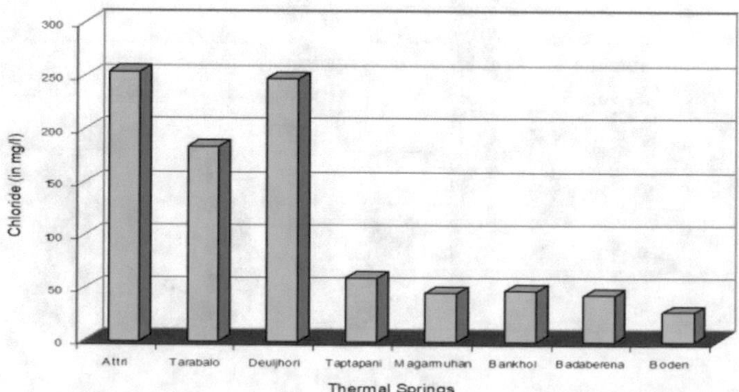

Fig. 5.9 Bar diagram showing variation of Cl Conc. of different thermal water

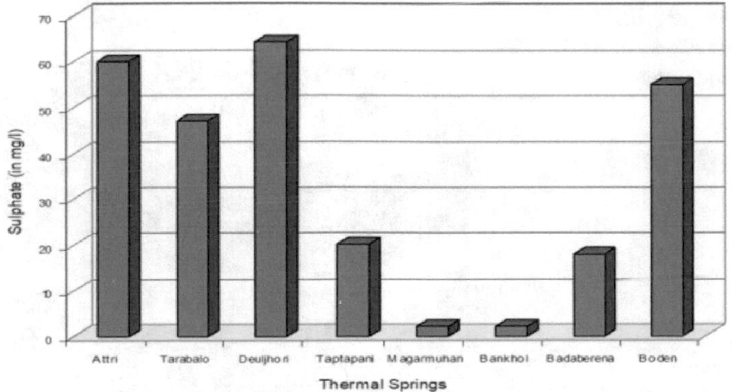

Fig. 5.10 Bar diagram showing variation of SO_4 Conc. of different thermal water

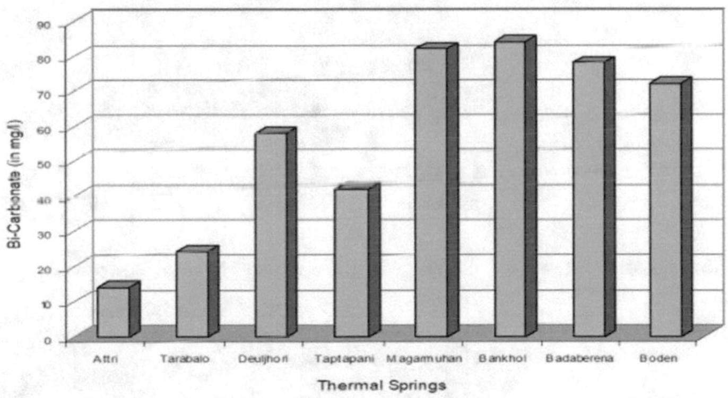

Fig. 5.11 Bar diagram showing variation of HCO_3 Conc. of different thermal water

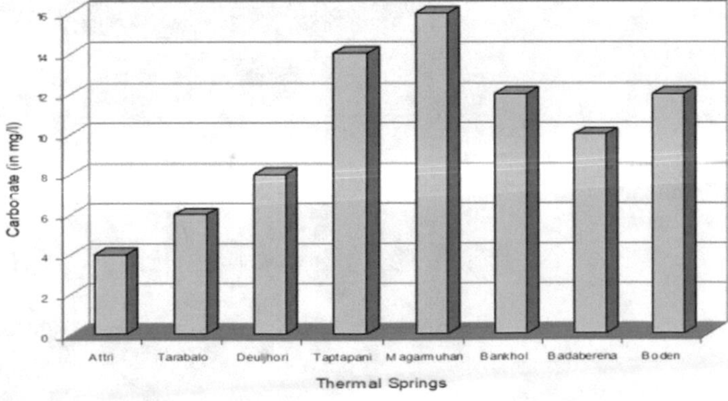

Fig. 5.12 Bar diagram showing variation of CO_3 Conc. of different thermal water

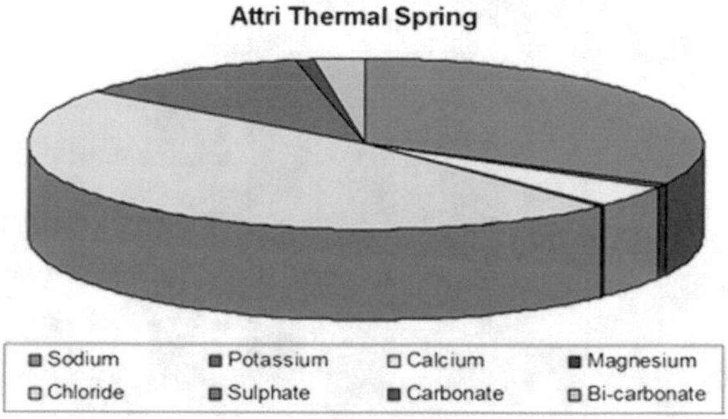

Attri Thermal Spring
Sodium
Potassium
Calcium
Magnesium
Chloride
Sulphate
Carbonate
Bi-carbonate

Fig. 5.13 Pie diagram showing Ionic Conc. of Attri thermal spring water

Tarabalo Thermal Spring
Sodium
Potassium
Calcium
Magnesium
Chloride
Sulphate
Carbonate
Bi-carbonate

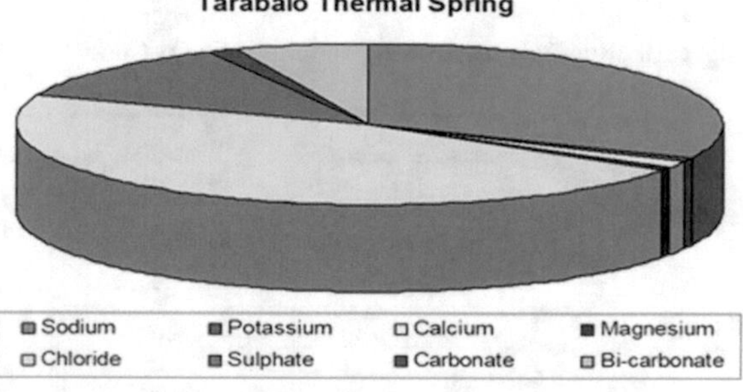

Fig. 5.14 Pie diagram showing Ionic Conc. of Tarabalo thermal spring water

Deuljhori Thermal Spring
Sodium
Potassium
Calcium
Magnesium
Chloride
Sulphate
Carbonate
Bi-carbonate

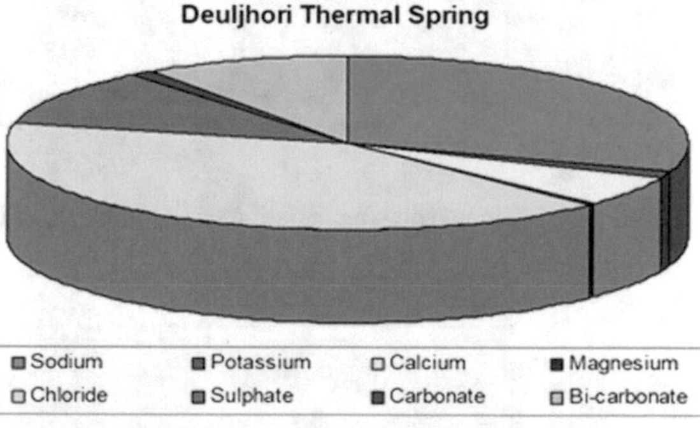

Fig. 5.15 Pie diagram showing Ionic Conc. of Deuljhori thermal spring water

Taptapani Thermal Spring
Sodium
Potassium
Calcium
Magnesium
Chloride
Sulphate
Carbonate
Bi-carbonate

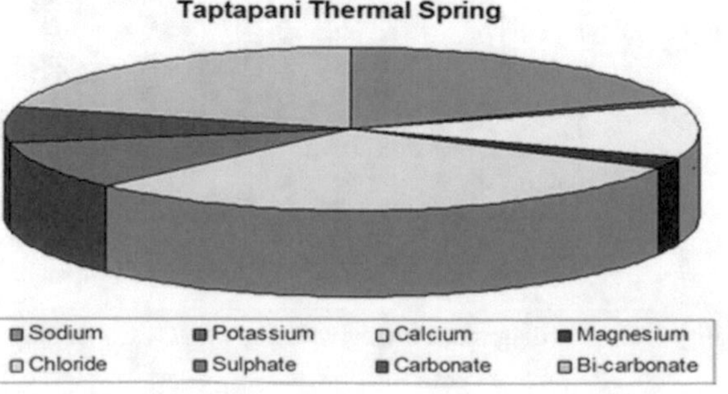

Fig. 5.16 Pie diagram showing Ionic Conc. of Tatapani thermal spring water

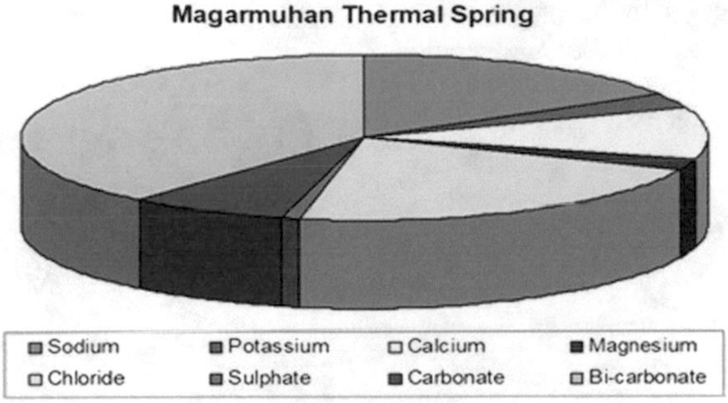

Fig. 5.17 Pie diagram showing Ionic Conc. of Magarmuhan thermal spring water

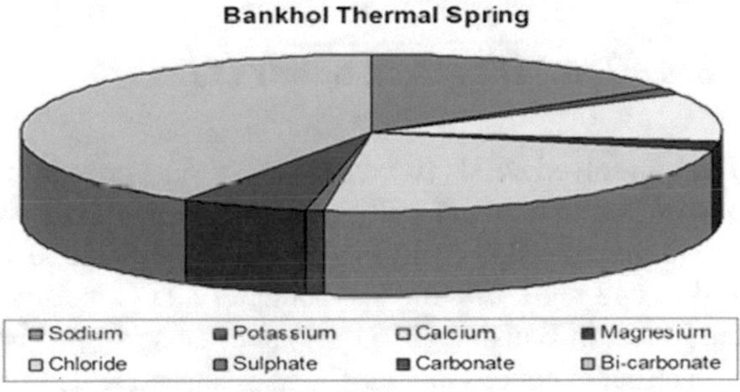

Fig. 5.18 Pie diagram showing Ionic Conc. of Bankhol thermal spring water

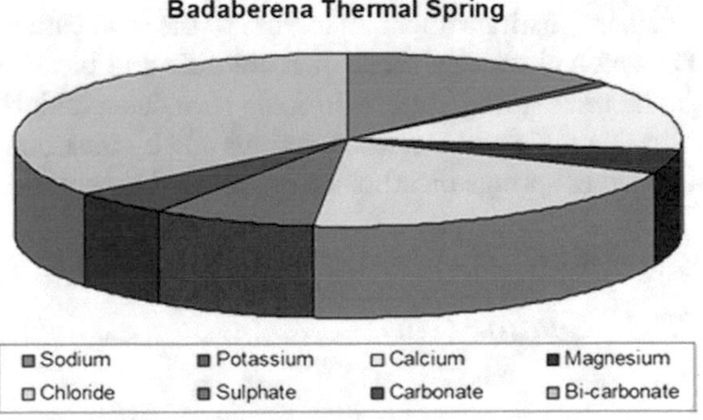

Fig. 5.19 Pie diagram showing Ionic Concentration of Badaberna thermal spring water

5.3.1 *Specific Conductance (EC)*

It is defined as the conductance of a cubic centimetre of water at a standard temperature of 25 °C (Todd 1980). The specific conductance (EC) of the thermal spring water was determined near the springs in the field itself. The value ranges from 345 to 1055 micro-mhos/cm. The highest specific conductance value was recorded in Deuljhori (1055 micro-mhos/cm) thermal spring and the lowest in Taptapani (345 micro-mhos/cm). This is shown graphically in Fig. 5.3.

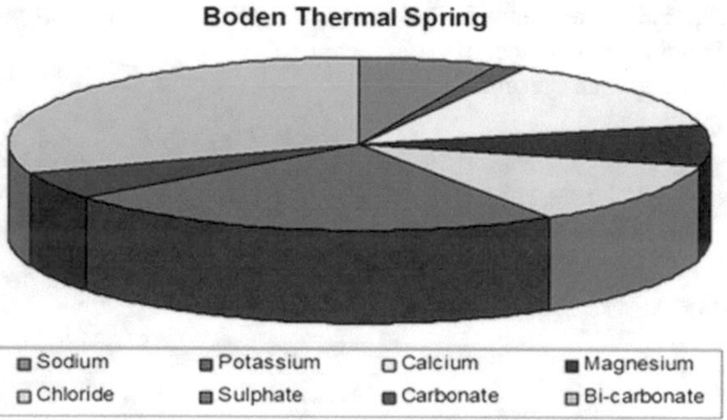

Fig. 5.20 Pie diagram showing Ionic Concentration of Boden thermal spring water

5.3.2 Total Dissolved Solids (TDS)

Total dissolved solids (total amount of salt) present in the thermal spring water are determined in terms of milligram per litre (mg/l) by weight of ions. Analysis results reveal that the TDS in all the thermal springs are low. The TDS value ranges from 214 to 642 mg/l. The highest content of TDS is found in Deuljhori (642 mg/l) and the lowest in Bankhol (216 mg/l) thermal springs (Fig. 5.4).

The major cations include Ca, Mg, Na, and K. The cationic chemistry is dominated by Na and Ca. In cationic abundance Na is followed by Ca, Mg, and K. Na and Ca together constitute about 70 to 96% of the total cations in the thermal spring waters. The anion chemistry shows that chloride and bicarbonate are the dominant ions in the thermal spring waters. In ionic abundance chloride is followed by sulphate in Attri, Tarabalo and Deuljhori springs and by bicarbonate in Taptapani spring, where as in other springs bicarbonate is followed by chloride.

5.3.3 Sodium (Na)

Among the cations sodium (Na) exists in much excess amount than other cations in most of the thermal springs of Odisha. Its concentration varies from 15 to 184 mg/l. The highest concentration is recorded in Deuljhori thermal spring (184 mg/l) whereas the lowest content is found in Boden (15 mg/l) spring (Figs. 5.5, 5.15 and 5.20).

5.3.4 Potassium (K)

Analysis results indicate that concentration of potassium (K) is much less in comparison to sodium (Na) in all the thermal spring water and the value ranges from 2 to 6 mg/l. Highest concentration of potassium is found in Deuljhori and Magarmuhan thermal springs and the lowest in Taptapani and Badaberena thermal springs (Figs. 5.6, 5.15, 5.17, 5.16 and 5.19).

5.3.5 *Calcium (Ca)*

Next to sodium, calcium occurs as the major constituent in all the thermal spring water. The concentration ranges from 6 to 35 mg/l. The highest value is recorded in Deuljhori (35 mg/l) spring whereas Tarabalo (6 mg/l) shows the lowest concentration (Figs. 5.7, 5.15 and 5.14). Low calcium content in the thermal water of Tarabalo is due to higher water temperature as calcium is more soluble at lower temperature than at higher temperature (Wright 1991).

5.3.6 *Magnesium (Mg)*

In all the thermal springs magnesium occurs as low in concentration. Its concentration varies from 2 to 18 mg/l. The highest value (18 mg/l) is recorded in Boden thermal spring (Fig. 5.8).

5.3.7 *Chloride (Cl)*

Among the anions chloride (Cl) occurs in much excess in amount in Attri, Tarabalo and Deuljhori spring water. The concentration of chloride varies from 28 to 254 mg/l. The highest chloride content (254 mg/l) is recorded in Attri thermal spring while the spring at Boden (28 mg/l) shows the lowest value (Figs. 5.9, 5.13 and 5.20).

5.3.8 *Sulphate (SO_4)*

Sulphate (SO_4) occurs in appreciable quantity in almost all the springs except Magarmuhan and Bankhol. The concentration values range from 2 to 64 mg/l. The highest concentration of Sulphate is found in Deuljhori (64 mg/l) and the lowest in Magarmuhan (2 mg/l) and Bankhol (2 mg/l) springs (Figs. 5.10, 5.15, 5.17 and 5.18).

5.3.9 Bicarbonate (HCO_3)

Next to chloride, bicarbonate (HCO_3) present in excess amount in all the springs. The concentration of bicarbonate varies from 14 to 84 mg/l. The highest value of bicarbonate is recorded in Bankhol (84 mg/l) and lowest in Attri hot spring (14 mg/l) (Figs. 5.11, 5.18 and 5.13).

5.3.10 Carbonate (CO_3)

Carbonate is recorded in much less amount in all the springs. The concentration of carbonate varies from 4 to 16 mg/l. The highest value of carbonate is noted at Magarmuhan (16 mg/l) and the lowest in Attri (4 mg/l) (Fig. 5.12, 5.17 and 5.13).

5.4 Type of Thermal Water

To understand the chemical behaviour of the thermal discharge, the chemical type of waters were determined based on the major ratios i.e., the concentration of anions and cations divided by molecular weight of cations and anions (Karantha 1987). The types of water so obtained are given in Table 5.2. The water of thermal springs of Odisha is mainly of three types i.e. Sodium Chloride (NaCl) type, Sodium Bicarbonate ($NaHCO_3$) type and Calcium Bicarbonate ($CaHCO_3$) type. The thermal spring water of Attri, Tarabalo, Deuljhori and Taptapani are Sodium Chloride type while that of Magarmuhan, Bankhol, Badaberena are Sodium Bicarbonate type and Boden is of Calcium Bicarbonate type (Fig. 5.21).

5.5 Graphical Representation of Chemical Analysis Data

Presentation of chemical analyses in graphical form makes understanding simpler and quicker. The chemical analytical results of these spring water are treated by most commonly used graphical representation methods like Stiff's diagram, Piper's trilinear diagram, U.S. Salinity diagram and Wilcox diagram etc.

5.5.1 Stiff's Diagram

Stiff's (1951) diagram uses four parallel horizontal axes and one vertical axis. The equivalent per million (epm) value of four cations i.e. Na, K, Ca, Mg are plotted on one side and four anions i.e. Cl, SO_4, HCO_3, CO_3 on the other side of the vertical axis. The vertex points of the polygon are connected to give a pattern whose shape is characteristic of a given type of water. The analysis data are plotted in Stiff's diagram and shown in Fig. 5.22.

5.5.2 *Piper Trilinear Diagram*

Piper trilinear diagram was developed by Piper in (1944) which shows the relative concentrations of major cations (Na, K, Ca, Mg) and anions (Cl, CO_3, HCO_3, SO_4). Piper trilinear diagram is most useful to understand chemical relationships among the waters. The chemical quality data of the thermal springs are used in Piper trilinear diagram for graphical analysis (Fig. 5.23). This diagram combines three distinct fields of plotting, two triangular fields at the bottom and one intervening diamond shaped field in the centre. In all the three triangular fields each vertex represents 100% of reacting values. In the lower left and right triangular fields, percentage of reacting values of cations and anions are plotted respectively. The diamond shaped field above the cation and anion triangles can be used to present both anion and cation groups as a percentage of the sample. The overall characteristic of the water is represented in the diamond shaped field by projecting the position of the plots in the triangular fields. Different types of water can be distinguished by the position their

Table 5.2 Chemical type of thermal water

Serial number	Name of the thermal spring	Molar ratios of cations in decreasing order of concentration	Molar ratios of anions in decreasing order of concentration	Chemical type of water
1	Attri	Na, Ca, K, Mg	Cl, SO_4, HCO_3, CO_3	NaCl
2	Badaberena	Na, Ca, Mg, K	HCO_3, Cl, SO_4, CO_3	$NaHCO_3$
3	Bankhol	Na, Ca, Mg, K	HCO_3, Cl, CO_3, SO_4	$NaHCO_3$
4	Magarmuhan	Na, Ca, Mg, K	HCO_3, Cl, CO_3, SO_4	$NaHCO_3$
5	Boden	Ca, Mg, Na, K	HCO_3, Cl, SO_4, CO_3	$CaHCO_3$
6	Deuljhori	Na, Ca, K, Mg	Cl, HCO_3, SO_4, CO_3	NaCl
7	Tarabalo	Na, Ca, K, Mg	Cl, SO_4, HCO_3, CO_3	NaCl
8	Taptapani	Na, Ca, Mg, K	Cl, HCO_3, CO_3, SO_4	NaCl

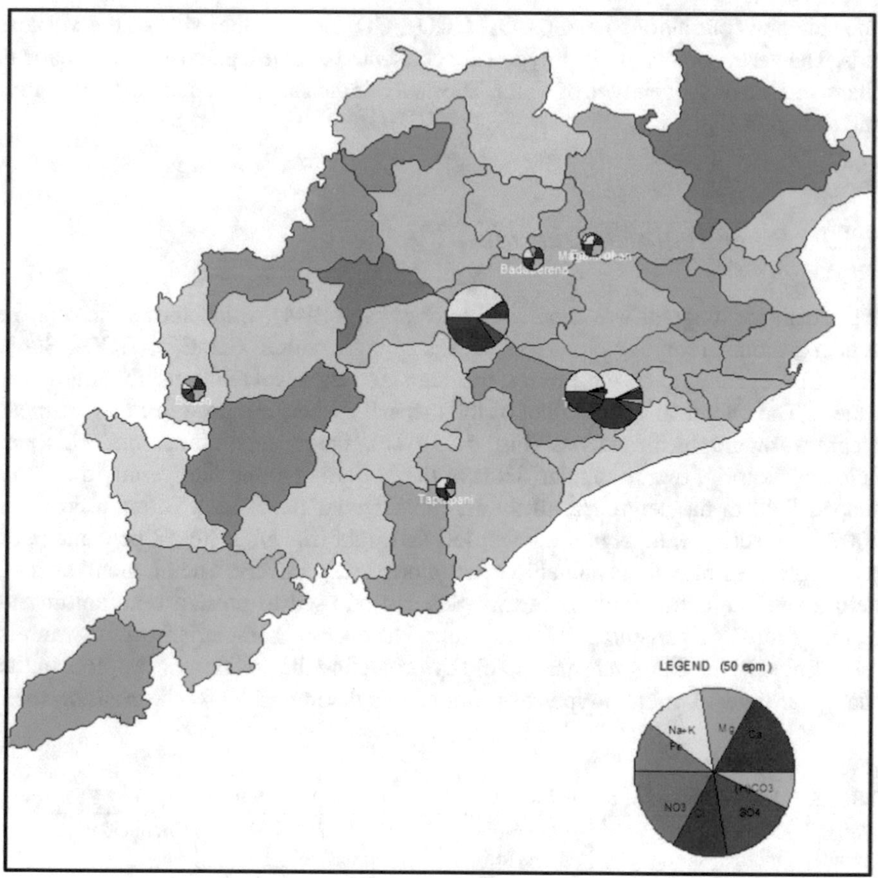

Fig. 5.21 Figure showing chemical composition of water in Pie diagrams of different thermal springs of Odisha

plotting occupy in certain sub areas of the diamond shaped field and their characters can be studied. The analysis value of cations and anions of all the thermal springs are plotted in Piper trilinear diagram (Fig. 5.23). The thermal springs at Attri, Tarabalo, Taptapani and Deuljhori fall in the area of the diamond shaped field which indicates that non-carbonate exceeds 50% i.e. chemical properties are dominated by alkalies and strong acids. Magarmuhan and Bankhol fall in the sub area which indicates that no one cation pair exceeds 50%.

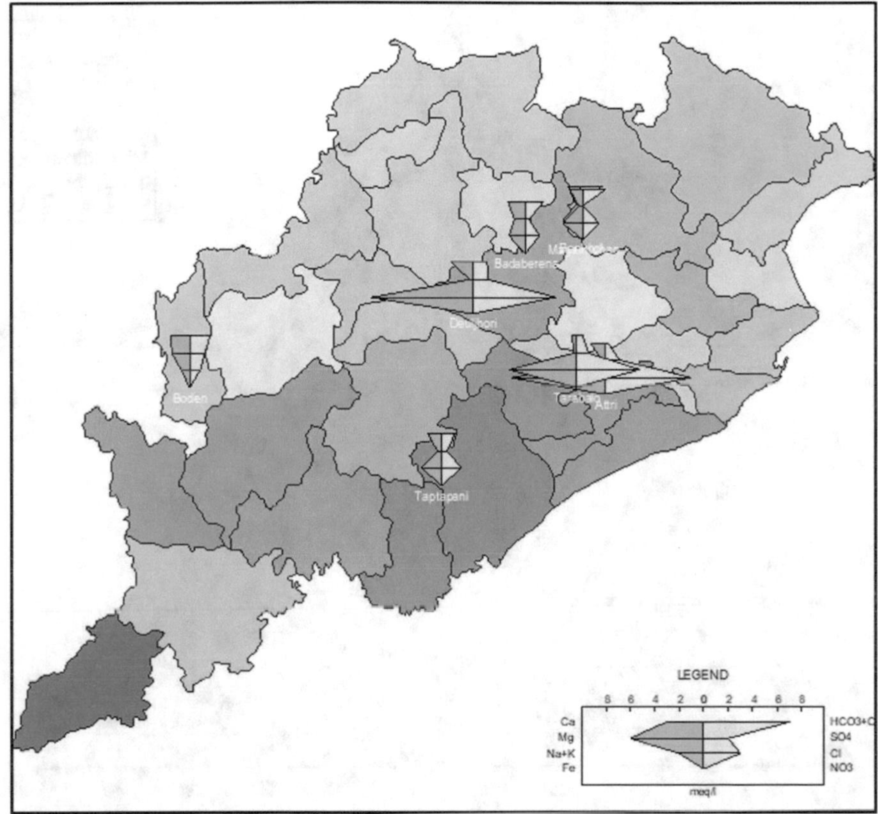

Fig. 5.22 Figure showing chemical composition of water in Stiff diagrams of different thermal springs of Odisha

5.5.3 US Salinity Diagram

This diagram was proposed by U.S. Salinity laboratory to determine the quality of water for irrigation use. This classification is based on the relation between specific conductivity (EC) and sodium absorption ratios (SAR). In the diagram SAR is taken on the ordinate and EC on the abscissa. The diagram shows twenty classes of water in terms of salinity hazard as low (C1), medium (C_2), high (C_3), very high (C_4), very very high (C_5) and also in terms of sodium hazard as low (S_1), medium (S_2), high (S_3) and very high (S_4).

The analysis results of thermal spring water are plotted on this diagram (Fig. 5.24). As per the classification, Attri, Tarabalo and Deuljhori belong to C_3S_2 class and all other springs fall in the C_2S_1 class.

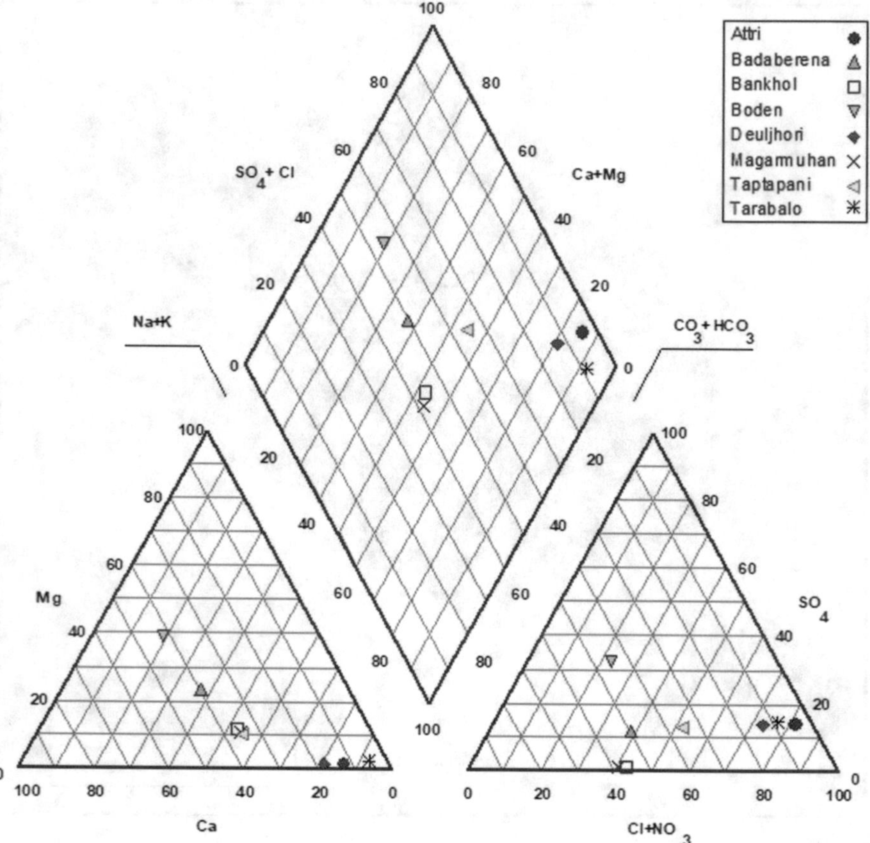

Fig. 5.23 Piper Trilinear diagram

5.6 Quality Assessment

The data obtained from chemical analysis are evaluated in terms of its suitability for drinking, other domestic and irrigation uses.

5.6.1 *Suitability for Drinking and General Domestic Use*

To assess the suitability for drinking and public health purposes, the hydro-chemical parameters of the thermal spring waters are compared (Table 5.3) with the prescribed specifications of WHO (1993) and Indian standard for drinking water i.e. ISO-10500:1991 (2003).

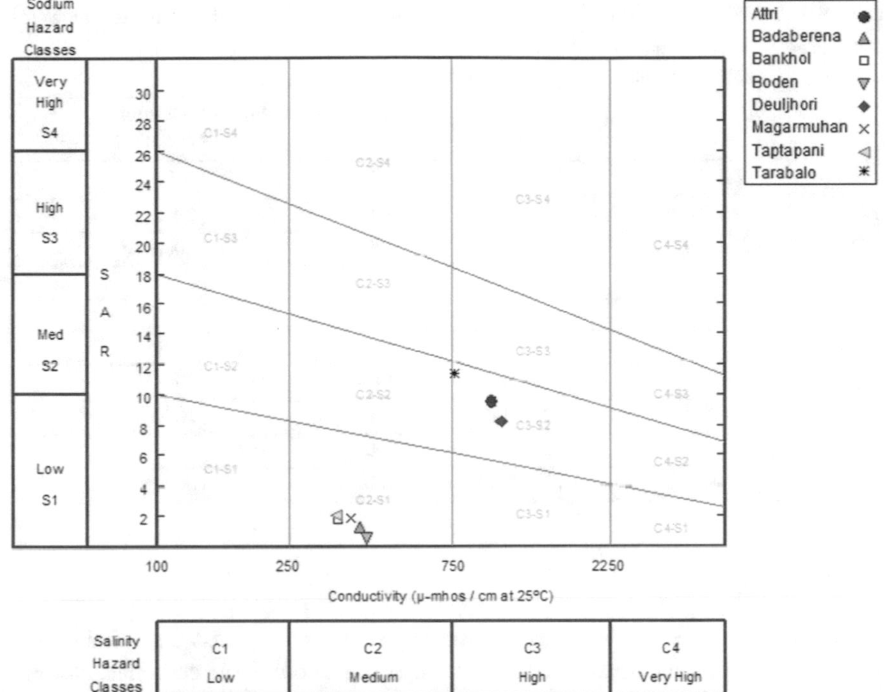

Fig. 5.24 US Salinity diagram

The specification values prescribed by WHO and Indian Standards indicate that most of the thermal spring waters are suitable for drinking and domestic uses as most of the parameters are within the permissible limits with few exceptions.

5.6.2 *Suitability for Agriculture Use*

Assessment of suitability of water for irrigation purpose requires consideration of TDS, concentration of any substance that may be toxic to plants and relative amount of certain constituents (Karantha 1987). Chemical constituents when present beyond certain limits in the water applied for irrigation may harm plant growth by toxicity or by changing soil properties. The parameters such as Sodium Absorption Ratio (SAR), Percent Sodium (Na%) are estimated to assess the suitability of water for irrigation purpose. EC and concentration of sodium are very important factors in classifying the irrigation water.

Table 5.3 Drinking water specifications (The concentrations are in mg/l. EC in micro-mhos/cm)

Parameters	Thermal spring WQ range (Odisha)	WHO (1993) specification	IS-10500:1991 specification	
			Highest desirable	Maximum permissible
pH	7.4–7.8	6.5–8.5	6.5	8.5
EC	345–1055	400-2000	500	2000
TDS	214–642	500–1000	500	2000
Na	15–184	20–175	–	–
K	2–6	10–12	–	–
Ca	6–32	100–200	75	200
Mg	2–18	30–50	30	100
Cl	28–254	25–600	250	1000
SO_4	2–64	25–250	200	400
HCO_3	14–84	–	300	600

Sodium Percentage (Na%)

Sodium content in water is a very important parameter to assess the suitability of water for agricultural purpose (Wilcox 1948). Sodium combining carbonate can lead to formation of alkaline soil while sodium combining with chloride form saline soils. Both these soils do not help in growth of plants. Further sodium by the process of Base Exchange replaces calcium in the soil, which reduces the permeability of the soil.

The percent sodium is calculated as

$$\text{Na\%} = \frac{\text{Na} + \text{K}}{\text{Na} + \text{K} + \text{Ca} + \text{Mg}} \times 100$$

(The concentrations are in meq/l).

The sodium percentage values thus calculated by applying the above equation for the thermal springs of Odisha are given in the Table 5.4. These values reveal that the thermal spring water of Attri, Tarabalo and Deuljhori has high sodium percentage.

The chemical quality of water samples of the thermal springs is studied from Na% vis-à-vis specific conductance in Wilcox diagram (Fig. 5.25).

Sodium Absorption Ratio (SAR)

The relative activity of Na ion in the exchange reaction with soil expressed in terms of a ratio known as Sodium Absorption Ratio (SAR), which is an important parameter for determination of suitability of water for irrigation use. Increase of sodium concentration in water effect deterioration of soil properties by reducing permeability (Kelley 1951; Tijani 1994). The sodium absorption ratio is calculated as:

Table 5.4 Sodium percentage (Na%) values of thermal springs of Odisha

Name of the thermal spring	Na% values
Attri	85.82
Tarabalo	92.27
Deuljhori	80.97
Taptapani	54.3
Magarmuhan	54.81
Bankhol	52.39
Badaberena	36.99
Boden	19.01

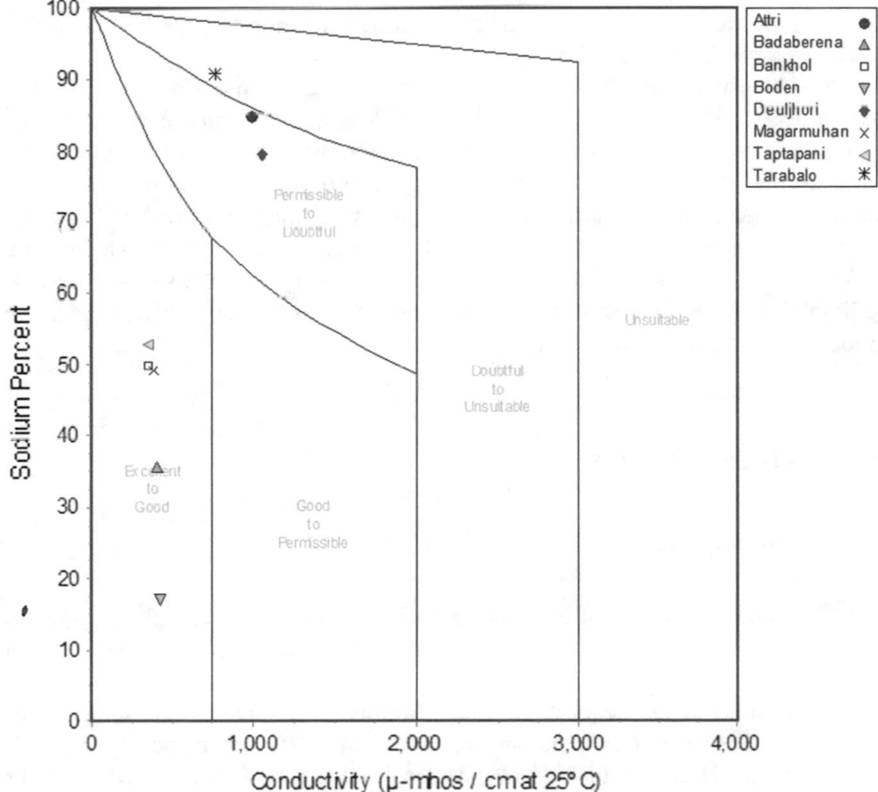

Fig. 5.25 Wilcox diagram

$$SAR = \frac{Na}{(Ca + Mg)/2}$$

where, the concentrations are in meq/l.

Table 5.5 SAR values of thermal springs of Odisha

Name of the thermal spring	SAR
Attri	9.508
Tarabalo	11.365
Deuljhori	8.174
Taptapani	1.951
Magarmuhan	1.797
Bankhol	1.704
Badaberena	1.193
Boden	0.524

The SAR values calculated by using the above equation for the thermal spring water are given in the Table 5.5.

The SAR value in these thermal spring water ranges from 0.524 to 11.365. The plot of data on US Salinity diagram where EC is taken as salinity hazard and SAR as alkalinity hazard shows that these spring water samples fall in the category C_2S_1 and C_3S_2 class (Fig. 5.24). The thermal spring water of Attri, Tarabalo and Deuljhori fall in C_3S_2 class that indicates a high salinity and medium sodium water. This water can be used on soils except soils with restricted drainage and fine texture having high cation exchange capacity. Rest of the springs belongs to medium salinity and low sodium class (C_2S_1). Their water can be used on almost all soils where moderate amount of leaching occurs.

5.7 Geothermometry

5.7.1 Introduction

The concentration of natural chemical constituents such as silica (SiO_2), sodium (Na), potassium (K), calcium (Ca), magnesium (Mg), aluminium (Al), iron (Fe) and manganese (Mn) in the thermal water are controlled by the particular temperature dependant mineral-water equilibria. The solubility of certain rock forming minerals and the existence of temperature dependant equilibria has helped to establish qualitative as well as quantitative geochemical indicators of sub-surface reservoir temperature. These indicators are called geothermometers (Fournier and Truesdell 1974; Ellis and Mahon 1977) and the process of estimating temperature of reservoir fluid with the help of geothermometers is called geothermometry.

The geothermometers are valuable tools in the evaluation of new geothermal fields. Geothermometers are of various types such as solute geothermometers, isotope geothermometers and gas geothermometers. Attention must be given to the following things to avoid common errors.

1. Data Quality: Sampling and analysis of water should be précised and accurate following standard procedures. An ionic balance is a useful check on the accuracy and completeness of analysis.
2. Spring Solution: Not all hot springs are equally valid for geothermometry and a priority should be established. Hot springs with high chloride concentration and good discharge provide most reliable result.
3. Host Rocks: Mineralogy of host rocks is important before applying geothermometers.
4. Well Discharge: Rapid flow of fluid to the surface overcomes some of the problems associated with hot spring water, but can introduce a few of their own.

5.7.2 Selection of Geothermometer

It is assumed to obtain similar estimates of reservoir temperatures from different geothermometers, but the reality is that geothermal systems are complex and many sub-surface processes can produce differences in geothermometry results.

Solute geothermometers are based on temperature dependant mineral fluid equilibria (Giggenbach 1981). The main quantitative chemical indicators of temperature are silica and Na/K ratio geothermometers. Fournier and Truesdell (1973) have developed another geothermometer, called Na–K–Ca geothermometer.

5.7.3 Silica Geothermometer

This geothermometer is based on the temperature dependence of the solubility of silica minerals in water. Silica occurs in various forms such as quartz, chalcedony, cristobalite and amorphous silica. Variation in solubility of silica with different value of temperature has been studied by various workers (Morey et al. 1962; Fournier and Rowe 1962; Arnorson 1975). For silica geothermometer a very useful graphical form was published by Fournier and Truesdell (1970). Later Truesdell (1976) expressed it through the following equation.

$$T_{SiO_2} = \frac{1315}{5.205 - \log_{10} S}$$

where T_{SiO_2} is silica reservoir base temperature in °C and S is dissolved silica in ppm.

5.7.4 Na–K Geothermometer

Na–K geothermometer has steadily evolved from the initial observation that the Na/K ratio in most thermal waters is controlled by equilibrium with albite and K-feldspar. Several workers derived geothermometry equations based on Na/K ratio. The most widely used equation given by Truesdell is

$$t^{o}C = \frac{856}{(\mathrm{Log}_{10}\mathrm{Na/K} + 0.857)} - 273$$

where, Na and K concentrations are in ppm.

The Na/K geothermometer has been successfully used for the prediction of reservoir temperatures for near neutral geothermal waters.

5.7.5 Na–K–Ca Geothermometer

It has been found that temperatures calculated from Na/K geothermometer are generally too high when applied to waters with high Ca. Considering the influence of calcium in alumino-silicate reaction, Fournier and Truesdell (1973) proposed the following empirical equation.

$$t^{o}C = 1647\left[\{\log_{10}(\mathrm{Na/K}) + \beta[\log_{10}\sqrt{\mathrm{Ca}}/\mathrm{Na} + 2.06] + 2.47\}\right] - 273$$

where $\beta = 4/3$ for $\sqrt{}$Ca/Na > 1

$\beta = 1/3$ for $\sqrt{}$Ca/Na < 1

and Na, K, Ca concentrations are in ppm.

5.7.6 Estimation of Reservoir Temperatures

The reservoir temperature is estimated by using Na/K and Na–K–Ca geothermometers (Table 5.6), as the silica concentration is not determined for the water samples of the thermal springs.

5.8 Discussion

Chemical analysis results indicate that the chemistry of thermal springs of Odisha is not identical to each other. The solute concentrations in geothermal fluid vary due to variation in temperature, rock type and fluid source. Varied lithology of the

Table 5.6 Reservoir temperatures of thermal springs of Odisha

Name of the spring	Reservoir temp in °C by using Na–K–Ca geothermometer	Reservoir temp in °C by using Na/K geothermometer
Attri	112	129
Tarabalo	251	277
Deuljhori	124	92
Taptapani	125	127
Magarmuhan	180	215
Bankhol	149	181
Badaberena	131	151
Boden	170	209

area facilitates the availability of major cation and anions that are dissolved in local meteoric water to generate various types of thermal water. Two major types of thermal water namely NaCl type and $NaHCO_3$ type except Boden, ($CaHCO_3$ type in nature) exist in Odisha. The pH values as shown in the table indicate a near neutral to mildly alkaline character of the geothermal water. The total dissolved solids content values reveal that the thermal springs of Odisha are low in mineral content.

The concentration of sodium and potassium is controlled by temperature dependant mineral fluid equilibrium, which forms the basis of Na–K geothermometry. The sodium content in Attri, Tarabalo and Deuljhori thermal springs is appreciably higher than other springs. The potassium content does not vary significantly and occur in less concentration in comparison to sodium in all the springs. Sodium and potassium concentrations in thermal spring water may be derived by the reaction of water with alkali feldspars present in gneisses, granites and pegmatites (Bandyopadhyaya and Nag 1989). Prevalence of sodium over potassium is due to greater solubility of sodium and greater tendency of potassium to be adsorbed in the altered wall rock. Calcium concentration is usually at low level in high temperature fluids and increases with acidity and salinity. The calcium concentration in all the springs does not vary greatly except Tarabalo thermal spring, which has the lowest calcium concentration. Magnesium concentrations in high temperature geothermal fluids are usually very low. Higher concentration indicates near surface reactions leaching magnesium from the local rocks or dilution by magnesium rich ground water. All the springs have lower concentration of magnesium except Badaberena and Boden, whose waters contain little excess of magnesium than those of other springs. Low calcium and magnesium content in spring waters indicate a very low hardness of the water. Calcium and magnesium are likely to be derived from reactions involving minerals like plagioclase feldspars, pyroxene and amphiboles (Feth et al. 1964).

Sulphate concentration is generally low in deep geothermal fluids. The sulphate content of Attri, Tarabalo, Deuljhori and Boden is relatively higher than those of other springs. Bicarbonate occurs appreciably in high concentration at Magarmuhan, Bankhol, Badaberena and Boden thermal springs. Chloride content in the thermal water may be derived from the country rock as average chloride content of granitic rocks is 330 ppm (Rankama et al. 1950). According to (Narula et al. 1996) chloride is also produced by water-rock interaction with granitic and schistose rocks. Sulphate content in spring waters may be derived from the sulphide minerals mostly pyrite present in the rocks near the thermal spring areas. The reservoir temperature is calculated by using Na/K and Na–K–Ca geothermometers.

References

APHA (1995) Standard methods for the examination of water and waste water, 19th edn. APHA, 1015, 15th street, Newyork, Washington DC. 20005

Arnorson S (1975) Application of silica geothermometer in low temperature hydrothermal areas in Iceland. Am J Sci 275:763

Bandyopadhyay G, Nag SK (1989) A study on hydrology and chemical characteristics of Bakreswar group of thermal springs, Birbhum district West Bengal. Ind J Geol 61(1):20–29

Bhargav JS, Srivastav SK, Naik PK (2001) A study on the geochemistry of Attri hot spring area, Orissa. Bhu-Jal, (CGWB) 16(324)

BIS (2003) Indian standard drinking water specification ISO-10500:1991. Edition 2.1. Bureau of Indian standards, New Delhi

Chaterjee GC (1969) Mineral and thermal waters of India. Publ. of 23rd Geological Congress, vol 19, pp 21–43

D'Amore, Truesdell JH (1979) Models of steam chemistry at Larderello and the geysers. In: Proceedings 5th workshop, Geothermal reservoir engineering, Stanford, California, pp 283–297

Deb S, Mukherjee AL (1969) On the genesis of few groups of thermal springs in Chhotnagpur gneissic complex India. J Geochem Soc India 4:1–9

Ellis AJ, Mahon WAJ (1977) Chemistry and geothermal systems. Academic press, New York

Feth JH, Roberson C, Polzer WL (1964) Sources of mineral constituents in water from granitic rocks, Siera Neveda. USGS water supply paper. 1535-I, pp 11–170

Fournier RO, Rowe JJ (1962) Estimation of underground temperatures from the silica content of water from hot springs and wet steam wells. Am J Sci 264:685–697

Fournier RO, Truesdell AH (1970) Chemical indicators of sub-surface temperatures applied to hot spring waters of Yellowstone national park, USA. Proc. UN Symp. On development and utilization of geothermal resources. Pissa, vol 2, pt-1, Geothermics Spl Issue 2, pp 529–535

Fournier RO, Truesdell AH (1973) An empirical Na–K–Ca geothermometer for natural waters. Geochem Cosmochem Acta 37, pp 1255–1275. Imalayan Geology vol 4, pp 492–515

Fournier RO, Truesdell AH (1974) Geochemical indicators of sub-surface temperature. U.S.G.S Journ Res 2(3):259–1275

Giggenbach WF (1981) Geothermal minerals equilibria. Geochim Cosmochim Acta 45:393–410

Gupta ML, Saxena VK (1979) Geochemistry of thermal waters of Konkan region—estimation of reservoir temperature and limitations. Geoviews VI(1–4):57–83

Gupta ML, Sukhija BS (1974) Preliminary studies of some geothermal areas in India. Geotermics 3:105–112

Gupta ML, Saxena VK, Sukhija BS (1975a) Analysis of the hot spring activity of Manikaran area, Himachal Pradesh, India by geochemical studies and tritium concentration of spring waters. Proceedings 2nd UN Symp. Development and use of geothermal resources, San Fransisco, US, govt. printing office, Washington, DC, pp 741–744

Gupta ML, Hari Narain, Saxena, VK (1975b) Geochemistry of thermal waters from various geothermal provinces of India. Publ. No. 119 of the International Association of Hydrological sciences, pp 47–58

Gupta ML, Singh SB, Rao GV (1976) Geochemistry of thermal waters from various geothermal provinces of India. Int Asso Hydrol Sci. Publ. 119, pp 47–58

Karantha KR (1987) Ground water assessment, Development and Management. TATA McGrawhill publishing company Ltd. New Delhi. pp 1–720

Kelley WP (1951) Alkali soils—their formation, properties and reclamation. Reinhold Publishing Corporation, New York

Morey GW, Fournier RO, Rowe JJ (1962) The solubility of quartz in water the temperature interval at from 29 to 300 °C: Geochem. et Cosmochim. Acta 26(10):1029–1043

Narula PL, Absar A, Pande P, Ravishanker (1996) Changes in chemistry of thermal waters consequent to the Uttarkashi earthquake, NW Himalaya. Geothermal energy in India, GSI spl publ. No-45, pp 209–222

Piper AM (1944) A graphical procedure in the geochemical interpretation of water analysis. Am Geophys Union Trans 25:914–923

Rankama K, Sahama Th G (1950) Geochemistry. Chicago University press, pp 912

Saxena VK (1983) Geochemical prospecting for the geothermal resources in Konkan coast and the Godavery valley, India. Ph.D. thesis, Rewa Unversity, India

Saxena VK, D'Amore F (1984) Aquifer chemistry of Puga and Chumatang high temperature geothermal systems in India. J Volcanol Geotherm Res 21:333–346

Saxena VK, Gupta ML (1982) Geochemistry of some thermal and cold waters of Godavari valley. J Geol Soc Ind 23(11):551–560

Saxena VK, Gupta ML (1985) Aquifer chemistry of thermal waters of the Godavari valley, India. J Volcanol Geotherm Res 25:181–191

Saxena VK, Gupta ML (1986) Geochemistry of thermal waters of Salbardi and Tatapani, India. Geothermics 15(5/6):705–714

Singh R, Bandyopadhyay AK (1995) Geochemical studies of some thermal springs in Hazaribagh district, Bihar, India. Indian Miner 49(1&2):55–60

Singh R, Kanwar SS, Jaggi GS, Kartha KNK (2004) Geochemistry of thermal springs from Bhutan Himalayas. J Geol Soc India 64:191–198

Stiff HA (1951) Interpretation of chemical water analysis by means of patterns. J Petrol V.3(10) Technical note 84, section 1, pp 15–16. sec-2, p.3. Dallas

Swain PK, Padhi RN (1986) Geothermal fields of Ganjam and Puri districts, Orissa. GSI, Records vol 114, Pt 3, pp 41–46

Tijani J (1994) Hydrochemical assessment of ground water in Maro area, Kwara state. Nigeria Envt geol 24:194–202

Todd DK (1980) Groundwater Hydrology. John Willey & sons. Singapore. pp 1–535

Truesdell AH (1976) Geotherm, A geothermic computer programme for hot springs system. In: Proceedings 2nd UN Symp. on developments and use of geothermal resources vol 1, pp 831–836

White DE (1970) Geochemistry applied to the discovery, evaluation and exploitation of geothermal energy resources. Geothermics 2(1):58–80

WHO (1993) Guide lines for drinking water quality vol 1. Recommendations, Geneva. WHO; pp 1–4

Wilcox LV (1948) The quality of water for irrigation use. U.S. Department of Agriculture, Tech Bull 1962, Washington D.C. m19p

Wright PM (1991) GHC Bulletin. pp 8–12

Chapter 6
Characteristics of Geothermal Gases

Abstract This chapter includes the type and nature of gaseous phases associated with the thermal springs of Odisha. The gases emanate along with water as bubbles causing micro rings of waves when reaches the water surface. The gases associated with thermal springs are collected in bottles by water displacement method at the point of its emergence and were analyzed for determination of gaseous composition. The constituents are nitrogen, oxygen, argon, helium, carbon dioxide and methane. While the nitrogen content is very high (88–90.5% by volume) the oxygen content is significantly low (1.2–6.6% by volume). Amount of helium, a rare gas is quite high in Attri (1.7%) in comparison to Deuljhori (0.768%) and Taptapani (0.269%). Methane is present as traces (0.5–0.7%). Argon content varies from 1.567% at Attri to 3.3% at Deuljhori. It is observed that percentage of nitrogen is high in thermal springs of Odisha and absence of equivalent oxygen may be due to circulation of meteoric water in the geothermal system. Sulphurous smell of the gases of thermal springs of Odisha is imparted due to possible alteration of sulphide minerals of the Easternghat rocks. Presence of helium is possibly due to disintegration of radioactive elements in Pre-cambrian terrain. Attempt at finding the causes of rise/fall of certain gas content from inside the springs have been made.

6.1 Introduction

Thermal springs are normally associated with various gas phases. The gases usually come out through the fissures from the sources of thermal springs in the form of bubbles in water bodies at different points and create micro-ring features on the water surface. Geochemical information about composition, temperature and pressure conditions of fluids within a geothermal system are generally derived from the chemical analyses of liquid phases reaching the surface in form of hot springs, mud pools, fumaroles etc. Knowledge of composition of gas phases associated with various surface emanations of a geothermal system, however, forms an essential part of the study of these systems. Information on qualitative and quantitative gas chemistry of a geothermal system helps in better understanding of the hydro-geochemical

S. C. Mahala, *Geology, Chemistry and Genesis of Thermal Springs of Odisha, India*, SpringerBriefs in Earth Sciences, https://doi.org/10.1007/978-3-319-90002-5_6

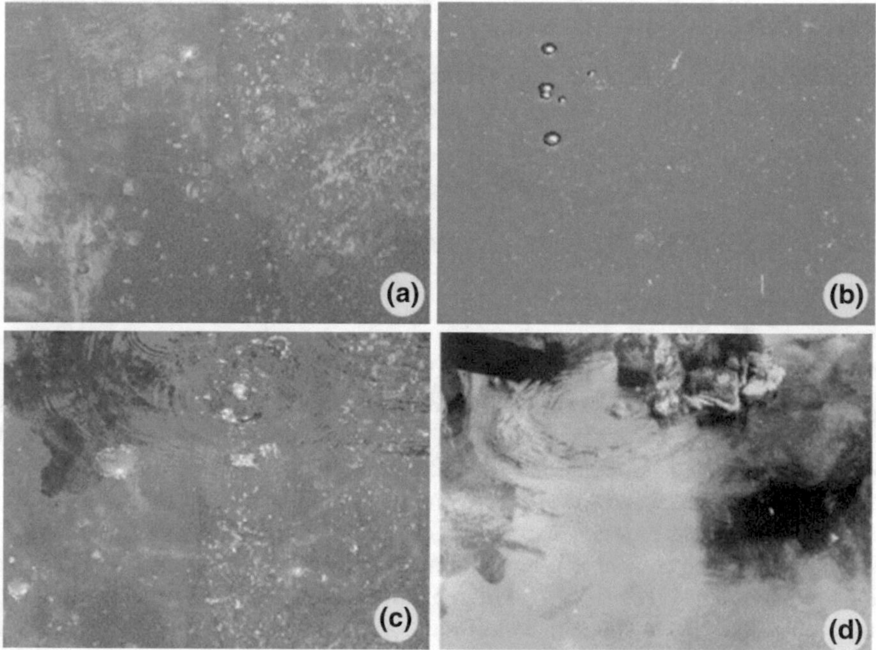

Fig. 6.1 Gas bubbles coming from thermal springs

model of the area and the conditions governing the fluid-mineral-gas equilibria in the geothermal system (Glover 1970). The most important parameter on which such equilibria depend is the temperature. Relative concentrations of major gases such as carbon dioxide (CO_2), hydrogen sulphide (H_2S), methane (CH_4), hydrogen (H_2) and nitrogen (N_2) are significant in this regard as they provide information on the thermal conditions in the deepest part by a geothermal system (Panichi et al. 1978). Gases undoubtedly play a major role in eruptive phenomena and their composition and relative abundance in steam phase are significant, because composition of gases coming out from a geothermal field can provide much more information and better understanding of the geothermal system. Considerable quantities of gas coming up can be observed as bubbles through the water at Attri, Taptapani and Deuljhori thermal springs areas (Fig. 6.1). Thermal spring at Attri has copious discharge of gas among all other springs.

6.2 Methods of Collection

Much work has been done in the past years in developing the techniques and methods for collection and analysis of geothermal gases depending on the nature, accessibility and other surrounding conditions of the geothermal field (Datta Munshi et al. 1984;

Saxena 1987). During the present study, gas samples from the thermal springs have been collected in gas jars by water displacement method. A plastic funnel of 30 cm diameter connected to the sampling tube fitted with the collecting jar was placed just above the point of steam/gas discharge at the bottom of the pool. The collecting bottles were filled with water and placed inside the water in an inverted position. The gas coming through the sampling tube came to the bottle, displaced the water and was stored inside. After collection, the gas jars were sealed immediately by sealing wax and kept inside a wooden box designed to keep the bottle safely in inverted position and taken to the laboratory. The samples were analyzed by means of Gas Chromatograph.

6.3 Results of Analysis

The results of analysis of gas samples are given in Table 6.1.

Analysis result reveals that nitrogen is the main component of geothermal gases (88–90.5%) followed by oxygen (1.2–6.6%) and they together constitute more than 95% in volume (Figs. 6.2 and 6.3). Other gases detected are helium, argon, traces of methane and rare carbon dioxide. Amount of helium is quite high in Attri (1.7%) in comparison to Deuljhori (0.768%) and Taptapani (0.269%) respectively (Fig. 6.4). Methane is present as traces (0.5–0.7%) (Fig. 6.6). Carbon dioxide is lacking in Attri and Deuljhori but very low (0.2%) at Taptapani. Argon content varies from 1.567% at Attri to 3.3% at Deuljhori (Fig. 6.5). The analytical results are also represented by pie diagrams (Figs. 6.7, 6.8 and 6.9). Composition of gases from various thermal springs in other parts of the country is given in the Table 6.2.

The table shows that helium content in gases of Bakreshwar thermal spring is high where it is extracted by applying proper technology. In this context the helium content of Attri thermal spring in Odisha (Table 6.1) seems to be feasible for economic exploitation.

Table 6.1 Composition of gases from thermal springs of Odisha

Name of the spring	N_2%	O_2%	He%	Ar%	CO_2%	CH_4%
Attri	90.54	1.2	1.7	1.56	–	0.5
Deuljhori	88.89	6.6	0.768	3.3	–	0.6
Taptapani	88	5.8	0.269	1.94	0.2	0.7

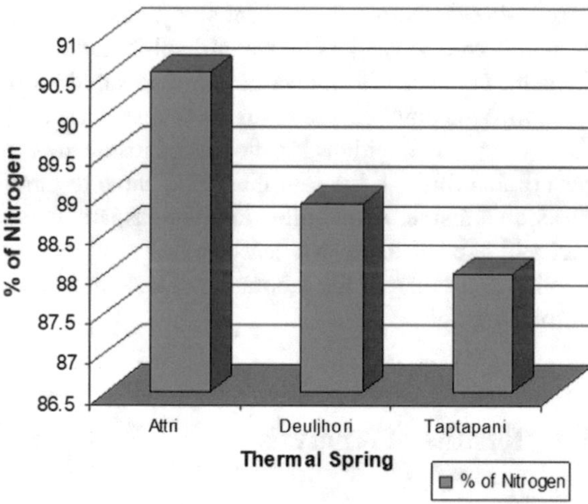

Fig. 6.2 Bar diagram showing Nitrogen%

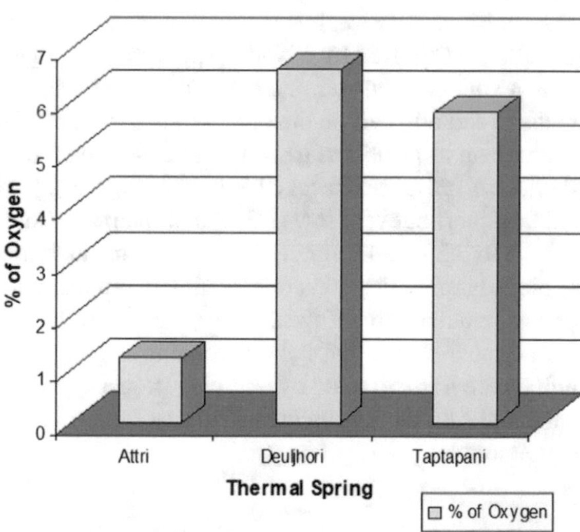

Fig. 6.3 Bar diagram showing Oxygen%

6.4 Discussion

Analytical results indicate that the nitrogen content is very high and the oxygen is significantly low in all the samples comparison to that of present atmospheric level (PAL). The composition of the present day atmospheric air is roughly 16% oxygen, 80% nitrogen with traces of inert gases (Datta Munshi et al. 1984). The gas composition of the thermal springs of Odisha, thus, differs significantly from the present atmospheric level (PAL). The geothermal systems can be classified on the basis of their gas composition (Saxena 1987). The thermal springs of Odisha can be classified as the nitrogen dominant type. In fact, the air saturated ground

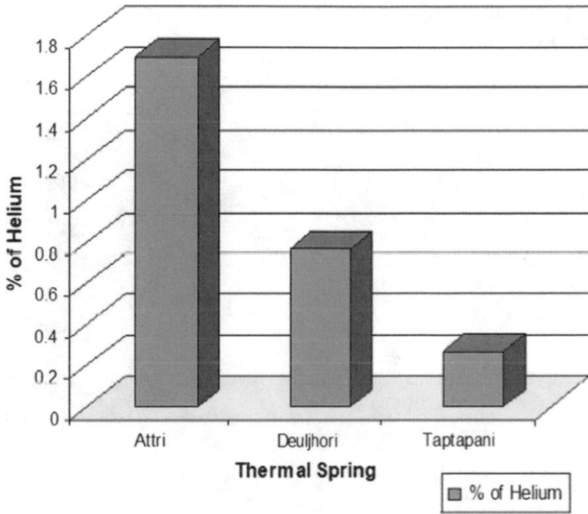

Fig. 6.4 Bar diagram showing Helium%

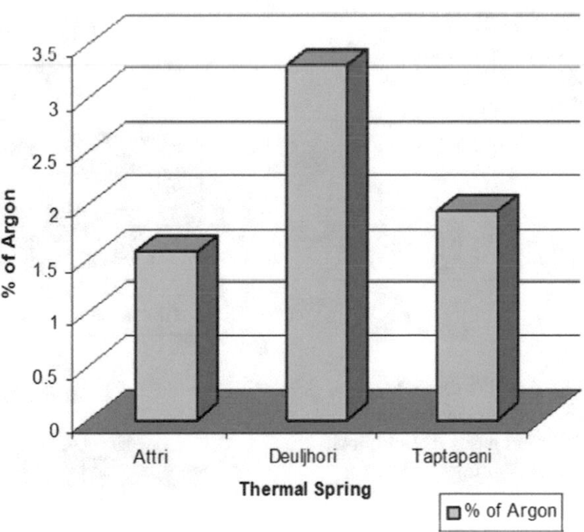

Fig. 6.5 Bar diagram showing Argon%

water is characterized by high N_2 (Giggenbach 1975; D'Amore and Nuti 1977). Comparable level of N_2 in geothermal gases suggests that N_2 has been derived from the atmosphere along with meteoric water (Saxena 1987). Besides, entrapped methane and ammonia of the primitive atmospheric air under high pressure and temperature might have got dissociated into nitrogen and hydrogen, which explains the high percentage of nitrogen in the thermal spring gases. Hydrogen produced inside the lithosphere gets rapidly utilized in the formation of juvenile water (Data Munshi et al. 1984). High CO_2 content is generally associated with geothermal system in volcanic and magmatic areas (White 1957; Arnorson and Barnes 1983). CO_2 may also be derived through thermal metamorphic process. Both magmatic and organic

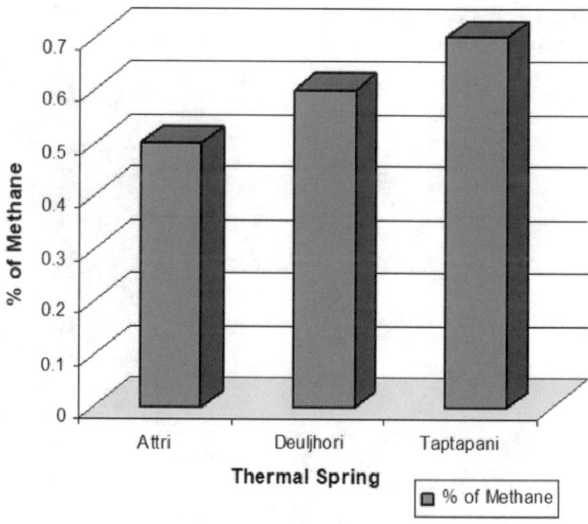

Fig. 6.6 Bar diagram showing Methane%

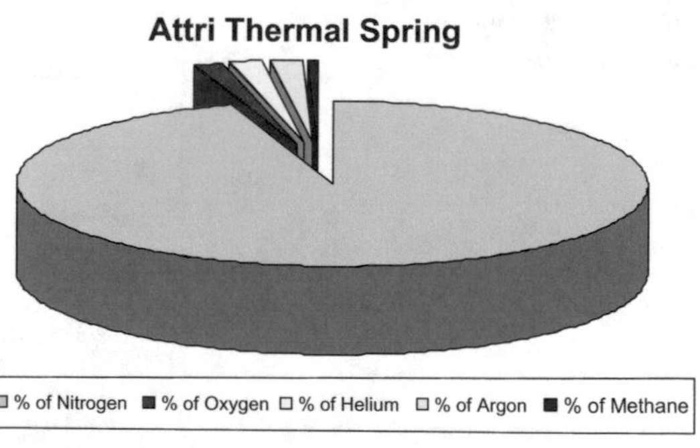

Fig. 6.7 Pie diagram showing concentration of different gases in Attri thermal spring

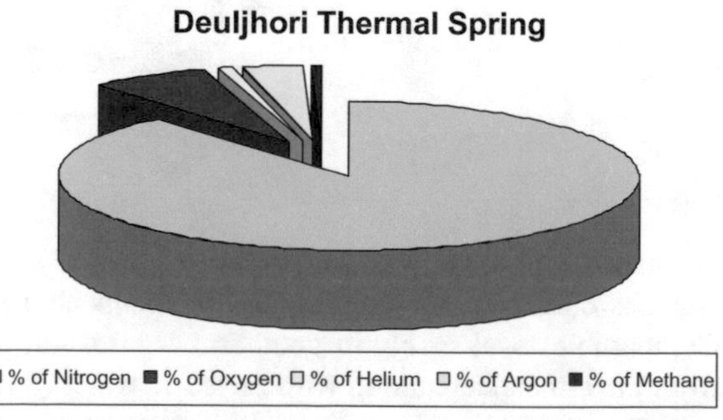

Fig. 6.8 Pie diagram showing concentration of different gases in Deuljhori thermal spring

origin has been put forth for the presence of CO_2 in the thermal springs (Ellis and Mahon 1977). Presence of traces of CO_2 in the thermal springs of Odisha, and their confinement to Pre-Cambrian terrain (Khondalite rocks associated with graphite) and absence of any recent volcanic activity around them lead to infer that it is of meteoric origin and derived through metamorphic and/or organic process. The thermal springs

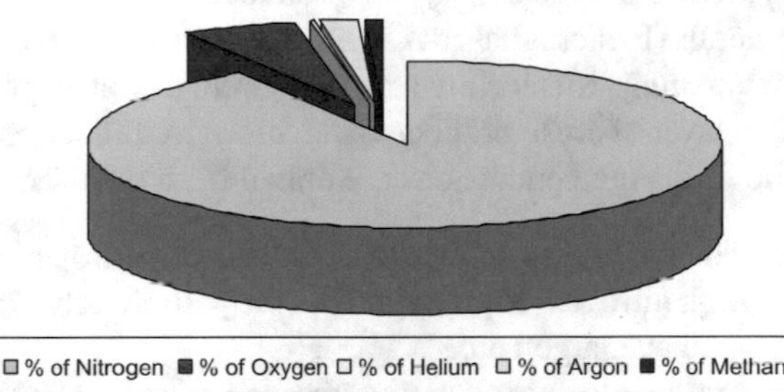

Fig. 6.9 Pie diagram showing concentration of different gases in Taptapani thermal spring

Table 6.2 Gaseous composition of different thermal springs in India (Conc. in vol.%)

Name of the thermal spring	N_2%	O_2%	He%	Ar%	CO_2%	CH_4%
Manikaran	0.7		0.002	0.02	98.4	0.075
Kasol	45.7	–	–	0.78	50.3	0.73
Sohna	86.24	0.28	0.002	1.04	10.51	0.13
Surajkunda	86.37	0.25	0.12	2.0	4.75	0.83
Tatapani	79.25	0.63	0.1	1.82	16.41	1.48
Tural	75.91	1.01	0.005	1.89	18.03	1.77
Rajawadi	75.5	1.2	0.01	2.14	18.39	1.42
Unhavere	74.6	0.96	0.01	0.25	20.66	1.21
Kloli	73.27	7.35	0.01	0.25	18.59	0.28
Agnigundla	73.98	1.04	0.07	2.08	19.45	2.19
Buga	79.54	2.49	0.001	0.74	15.81	0.25
Bhuttayagudem	71.48	2.14	0.001	0.73	24.91	0.59
Tantloi	85	1.8	0.6	1.3	–	1
Bakreshwar	88.87	–	2.64	–	3.61	4.88

Saxena (1987), Datta Munshi et al. (1984)

of Odisha are confined along major lineaments, which may be major faults or fissures (Singh et al. 1995). These structural features are considered as the avenue along which the atmospheric air gets entrapped in the lithosphere. The atmospheric air containing about 16% oxygen seems to be utilized in the oxidation reactions in the substrata and juvenile water is formed through synthesis of oxygen and hydrogen and probably therefore, the thermal spring gases exhibit lesser oxygen content. In fact, some hydrogen sulphide may be lost from steam rising in natural channels and near surface reactions of steam with organic material may create additional methane, ammonia and carbon dioxide in the steam.

The rare gas specially helium is very likely formed due to radioactive disintegration going on inside the earth's crust (Singh 1989). Presence of methane may be attributed to decomposition of organic matter inside the earth's crust. Sulphide mineralisation is recorded in Easternghat complex by Acharya et al. (1995). Hydrogen sulphide may be produced by heating of rocks containing sulphide minerals (Easternghat rocks) and organic matter, which imparts sulphur smell to the surrounding. Meteoric water that circulates at depth when comes in contact with deep layer of crust, produces heat through conduction may not be a wild speculation. The following conclusions are drawn from the aforesaid discussion.

1. The thermal spring gases of Odisha are nitrogen dominant type. They have very high nitrogen, low oxygen but proportionately high helium at Attri and traces of methane, argon etc.
2. Presence of nitrogen and absence of equivalent oxygen in all the gas samples indicate that dissolved air in meteoric water circulates underground during the process of a geothermal reservoir.
3. Hydrogen sulphide is possibly produced due to heating of sulphide minerals present in Easternghat rocks. This may also be produced under a reducing environment by decay of organic matter containing sulphur. Plenty of pyrite and pyrhotite are observed in the structurally weak zones specially of charnockite around Attri and Khurda.
4. Presence of faults/fractures facilitates rock-water interaction in the substrata at an augmented temperature.
5. It is speculated that the meteoric water may be percolated at depth and while coming in contact with deeper layer of crust, heat is absorbed through the process of conduction.

References

Acharya BC, Dash B, Rao DS, Sahoo RK (1995) Sulphide mineralization in parts of Easternghats complex, Orissa. Vistas in Geol Res Spl Publ No-1, UU:22–26

Arnorson S, Barnes I (1983) The nature of carbon dioxide water in Snaefellsnes, Western Iceland. Geothermics 12(2/3):171–177

D'Amore F, Nuti S (1977) Note on chemistry of geothermal gases. Geothermics 6:39–45

Datta Munshi JS, Billgrami SK, Verma PK, Datta Munshi J, Yadav RN (1984) Gases from thermal springs of Bihar, India. National Academy "Science letters" India 7(9):291–293

Ellis AJ, Mahon WAJ (1977) Chemistry and geothermal systems. Academic press, New York

Glover RB (1970) Geothermics. Spl Issue 2(2):1355–1356

Giggenbach WF (1975) A simple method for the collection and analysis of volcanic gas samples. Bull Volc 39:132–145

Panichi C, Nuti S, Noto P (1978) Remarks on the use of the isotope geothermometers in the Larderello geothermal field. In: Proceeding IAEA symposium nuclear technology in hydrology, Munich

Saxena VK (1987) Application of gas and water chemistry to various geothermal systems of India. J Geol Soc India 29:510–517

Singh JR (1989) Chemistry of geothermal gases: collection, analysis and some Indian case studies. Ind Miner 43(1):7–18

Singh P, Das M, Ray P, Mahala S (1995) Thermal springs of Orissa: An appraisal. Utkal University, Spl Publ 1:244–254

White DE (1957) Magmatic, connte and metamorphic water. Geol Soc Am 68:1659–1682

Chapter 7
Genesis of Thermal Springs

Abstract In this chapter an attempt has been made to infer about origin of heat, water, gas and mineral constituents related to thermal springs of Odisha and to correlate them with the genesis of thermal springs. The source of heat is considered as the geothermal gradient inside the earth, from fault zones during tectonic movement in the crust, chemical changes (exothermic reactions) and disintegration of radioactive elements. The fluid phase may be derived from percolation of meteoric water inside the crust through structural openings which is subsequently heated up and manifest in the form of hot springs on the surface of the earth. Water during circulation may be charged with atmospheric gases and this may be the cause of origin of nitrogen and oxygen etc. The gases such as methane and carbon dioxide are formed due essentially to decomposition of organic materials. The rare gas helium, associated with the thermal springs, is regarded to be originated by the disintegration of radioactive minerals. The mineral constituents present in the spring water are due to the rock assemblages through which the hot meteoric water has been circulated. The chemical constituents are formed due to interaction between hot meteoric water and the mineralogical composition of the rocks. The chemical components present in the thermal springs of Odisha have not only enabled to categorize the type of water but also impart therapeutic value.

7.1 Introduction

Origin of thermal springs deals with source of heat, water, gas and mineral constituents. As already mentioned, a hot spring is a spring that is manifested by the emergence of geothermally heated water from the earth's interior. There exist numerous hot springs throughout the world, on every continent. Several views regarding the origin of the hot springs have been proposed so far by many workers. Amongst them, Allen and Day (1935), Day (1939), Barth (1950) and Ellis and Mahon (1977) have made valuable contributions regarding the origin of hot springs.

There exist two schools of thought expressing the origin of hot springs. Scientists from one school have correlated the hot springs with the volcanic activity or

S. C. Mahala, *Geology, Chemistry and Genesis of Thermal Springs of Odisha, India*, SpringerBriefs in Earth Sciences, https://doi.org/10.1007/978-3-319-90002-5_7

molten magma at depth in specific areas. They are of the opinion that large amount of circulating ground water are mixed up and interacted with the components from magmatic sources. The final products of such combination are the sources of mineral constituents and heated water of the hot springs. The other school of thought advocated for non-volcanogenic origin. They state that in such areas where there is lack of any volcanism or magmatic activity the heating of circulating ground water at depth may be caused due to geothermal gradient, disintegration of radioactive substances and exothermic reactions.

Attempt has been made to discuss the origin of thermal springs of Odisha on the basis of data documented during the present study.

7.2 Source of Water

Water of the thermal spring may be of magmatic or meteoric origin as surmised from various evidences like geological set up, structural disposition, tectonic system, hydrological characteristics and chemical nature. The meteoric water percolating down the surface dilutes the ascending juvenile fluid from depth. In a simple meteoric origin, the meteoric water percolating downwards along deep fractures may be heated by the rise of temperature under normal geothermal gradient. Lindgren (1935) has shown a rise of 1 °C temperature for every 30 m of vertical depth. Chaterjee and Guha (1968) considered that meteoric water is the source of the thermal springs, a statement that finds congruity by Deb and Mukherjee (1969). Chemical characters of water from the thermal spring sometimes yield evidences regarding its origin (White 1957b; Lovering and Morris 1965).

7.3 Source of Heat

Heat is accepted as an essential ingredient of the thermal springs. As a matter of fact the water issuing from a thermal spring is heated by geothermal gradient, i.e. heat from the earth's interior. The increase in rate of temperature with depth is known as geothermal gradient. In general, the temperature of rocks within the earth increases with depth. The meteoric water percolating deep down into the crust through structural breaks is heated up while coming in contact with the hot rocks inside. The water from thermal springs, in non-volcanic areas, is supposed to be heated in this manner.

In areas of volcanic activity water may be heated by coming in contact with magma. The high temperature magma may cause water to be heated enough so that it will start boiling or become superheated under pressure. The water becomes so hot that it builds steam with pressure. These erupt in a jet form on the surface termed as geyser (Yellowstone park of USA).

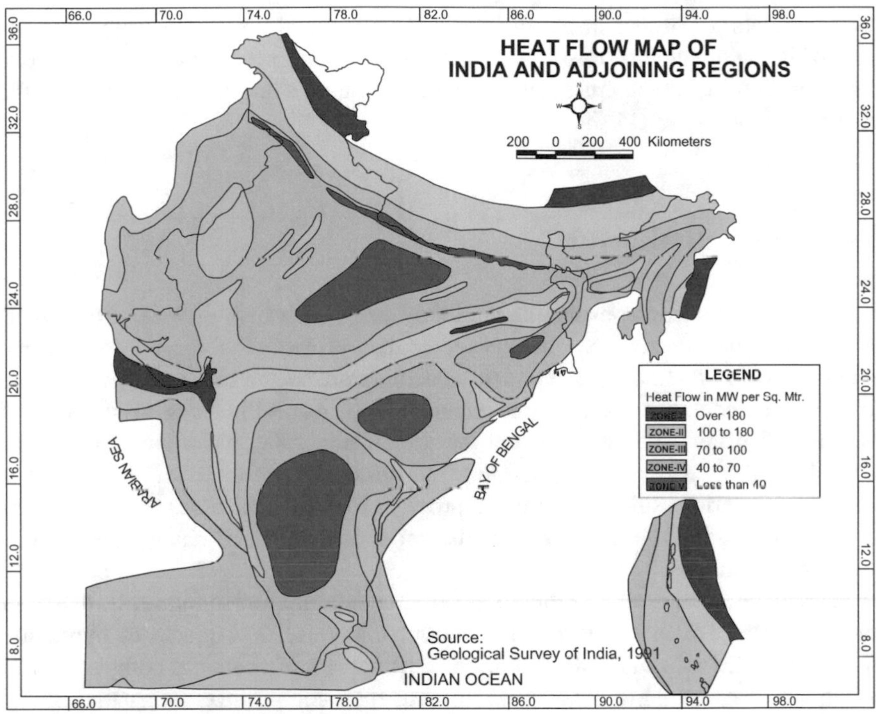

Fig. 7.1 Heat flow map of India & adjoining regions

But normal regional heat flow is considered as the source of heat for the hot springs in Indian shield. The heat flow, reflective of crustal thermal conditions, has been evaluated at a number of localities in Indian shield on the basis of the geothermal data and thermal conductivity data over corresponding depth sections. Heat flow zones of India have been determined and heat flow values range between 30 and 468 mW/m^2. According to Ravishanker et al. (1991), there are five heat flow zones in Indian sub-continent (Fig. 7.1). Odisha falls in zone-II (100–180 mW/m^2) and zone-III (70–100 mW/m^2) on the basis of heat flow values. Northwest of Easternghats in Odisha sector, Talcher coal field and thermal spring belt of Attri and Tarabalo fall within zone-II. The average temperature gradient is 61 ± 20 °C in this zone. Roy and Rao (1996) state that there is a variation in surface heat flow on the basis of analysis of U, Th and K in rocks of Precambrian crystalline terrains and basements of Gondwana basins. They further mentioned that variation in heat flow is primarily due to the heat generated by radioactive elements present in the crust.

In case of the thermal springs of Odisha the source of heat is a matter of speculation as very little data are available so far. Magmatic source of heat is quite remote as there is no known recent or sub-recent igneous activity in and around the thermal spring region. As already mentioned the thermal springs are mostly confined to tectonic zones in Pre-Cambrian terrain, it is possible that the spring water is heated due to deep circulation of water inside the earth. It is assumed that exothermic reactions e.g. oxidation of sulphide to sulphate, formation of carbonates and also radioactive disintegration inside the crust might have contributed some amount of heat in the geothermal system of Odisha.

7.4 Source of Minerals

The dissolved mineral constituents are likely to be derivative of water- rock interactions. It is the mineral composition rather than of rocks that has contributed the elements to the thermal spring water. The mineralized thermal spring waters are further enriched with certain chemical constituents in specific areas. So it is termed as 'Mineral water' and has imparted the therapeutic value. Dissolved CO_2 in the water has played a key role in such reaction through liberation of H^+ ions that has reacted with various silicates of the wall rock to release the alkalies, alkaline earths as well as chloride. As a result the water becomes depleted in H^+ resulting in alkaline nature of water.

Prevalence of Na over K is due to the greater solubility of sodium salts and their tendency to be derived from reactions involving minerals like plagioclases, pyroxenes and amphiboles in a high pH condition. Differences in chemical composition of waters are due mainly to differences in local rock composition rather than to the origin of the heat (Ellis and Mahon 1977).

7.5 Source of Gas

Thermal springs are normally associated with various gas phases. The gases come out through the fissures in rocks, from their sources along with spring waters in the form of bubbles that create micro-rings of waves on the water surface. Air saturated ground water is characterized by high nitrogen (90%) content. Considerable level of nitrogen in the geothermal gases suggests its derivation from the atmosphere along with meteoric water. Besides, entrapped methane and ammonia of the primitive atmospheric air under high pressure and temperature might have got dissociated into nitrogen and hydrogen which also explains high percentage of nitrogen in the thermal spring gases. Carbon Dioxide is generally associated with geothermal systems in the volcanic and magmatic areas. The thermal spring gases of Odisha having traces of carbon dioxide indicate a non-magmatic origin of these springs. The atmospheric air containing about 16% oxygen seems to be utilized in the oxidation process (Dutta Munshi et al. 1984). Therefore, the thermal spring gases probably exhibit presence of less amount of oxygen (6.6%). The rare gas such as helium is very likely formed due to radioactive disintegration and methane may be attributed to the decomposition of organic matter inside earth's crust (Ellis and Mahon 1977). Sulphurous smell of the thermal springs is obviously due to formation of hydrogen sulphide by heating of sulphide mineral-bearing rocks and/or organic matter.

7.6 Odisha Scenario

Thermal springs of Odisha are mainly confined to the crystalline schists and gneissic terrains of Precambrian age. Most of them are located in Precambrian terrain or close to the boundary of the Precambrian crystallines and Gondwana sedimentary rocks. These springs emerge more or less along a deep fault or fissure. Field studies supplemented by satellite data indicate that the thermal springs lie along lineaments, which infer close relationship between tectonism and thermal activity. There is no evidence of recent magmatic activity around these thermal springs. It is speculated that meteoric water percolating through faults and fissures may be circulated at depth and get heated while coming in contact with hot rocks. The mineral composition and radioactive character of the water is being influenced by the rock types through which the water has been circulated at higher temperature. Thermal systems associated with volcanism usually have very high TDS (White 1957a). As the water of the thermal springs in Odisha has low TDS, the heat is inferred to be derived from a non-magmatic source. The physico-chemical characters of the thermal water suggest that it is of meteoric origin. The normal geothermal gradient has great influence and hence, has been responsible for the rise of temperature of meteoric water. However, the source of heat may be derived from disintegration of radioactive elements and by exothermic reactions during metamorphism as well. According to Ghosh (1954) the Pre-Cambrian terrain is rich in radioactive elements. The quartz-pebble-conglomerate (QPC) of Iron Ore Supergroup also reported to have radioactive elements (Ray et al. 1990). Besides, presence of nitrogen and absence of equivalent oxygen in all the gas samples indicate that dissolved air in the meteoric water circulates underground during the process of recharging of a geothermal reservoir. It has already been mentioned that rocks of Easternghat Supergroup contain sulphide bearing minerals like pyrite, pyrhotite etc. These minerals, while coming in contact with hot water, impart sulphurous odour. As cited by Acharya (1966) Gautier's suggestion of foundering of sialic blocks and consequent squeezing out of water from the mass resulting in the formation of water vapour seems more plausible. This is because of the down faulting of Gondwana basin in foundering of the total infrastructure. The rough parallelism of the occurrences of the thermal springs with the edge of Gondwana basins lends support to this view. A conceptual model on the genesis of thermal springs of Odisha along Gondwana graben is given in (Fig. 7.2).

7.7 Discussion

Genesis of thermal spring is an important aspect of present investigation. It is a natural spectacular event associated with water, heat and gas components. The paradigm on the genetic aspect of these components has been drawn on the basis of database gen-

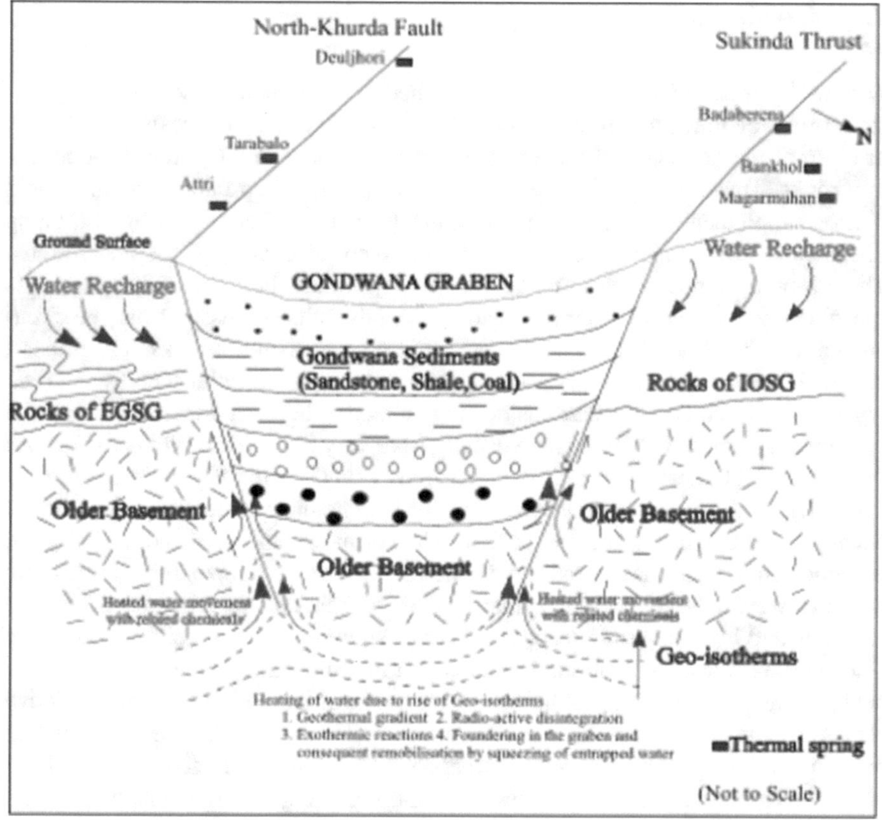

Fig. 7.2 Schematic diagram showing a conceptual model on the genesis of thermal springs of Odisha along Gondwana graben

erated during present study. Various parameters such as hydrological and chemical characteristics of the water, geological milieu and mineralogical—chemical constituents have been considered in order to infer the origin of the thermal springs in Odisha.

As discussed earlier the thermal springs manifest in various litho-assemblages belonging to Precambrian age and occurred in tectonic zones. They are issued along mega—lineaments and fault zones and majority of the springs are clustered along the Gondwana graben. There is no evidence of recent volcanism in and around the spring areas. So it is inferred that meteoric water has been considered as the main source of water of these thermal springs.

Hydrological parameters of the thermal spring water such as odour, pH and taste depend on the rocks through which it has been percolated. The higher temperature of the spring water is because of rise of temperature by geothermal gradient, exothermic reactions, and disintegration of radio-active elements. As there is no report of igneous activity in and around these areas, role of volcanism has not been considered for the temperature rise in these thermal springs. In fact the geothermal gradient has been responsible to add heat vis-a—vis rise of temperature of the spring water. These springs are invariably traced in metamorphic terrain; hence, metamorphic reactions (exothermic) involve release of heat that is also partly contributed to the rise of temperature of the water. The Pre-Cambrian metamorphites are the storehouse of radioactive minerals. Disintegration of these elements is also related to release of heat. Since these springs have been issued through the Pre-Cambrian terrain, they might have received heat from reaction of the radon. Thus, high temperature of these thermal springs has been inferred through multiple processes.

Chemical nature of the thermal spring water depends on presence of cations and anions. Chemical analysis data reveals that the thermal springs of Odisha are categorized as NaCl, $NaHCO_3$, and CaHCO3 types. The hot water percolates through various litho- types. The mineralogical assemblages of different rocks have played significant role to impart chemical constituents. The chemical elements present in different minerals have been dissolved in the hot water while coming in contact during circulation. In fact, it is the inter-action between rocks and hot water that imparts chemical nature of different thermal springs.

The gas phase associated with the thermal springs is another important topic of investigation. The gases are well noted by frequent discharge through bubbles. Analysis results show presence of gases such as nitrogen, oxygen, argon, methane and carbon dioxide in decreasing order (Table 6.1). Helium, a noble gas has been detected from different thermal springs. The rare gas helium is quite promising in the thermal springs of Odisha. It is detected as the highest percentage at Attri (1.7%). The gas phase of the thermal springs is different from the present atmospheric level (PAL). The higher percentage of nitrogen in the springs is probably due to the air saturated meteoric water and partly may be due to dissociation of ammonia/methane (primitive atmosphere). Nearly absence or trace of carbon dioxide can be explained by the absence of any recent magmatic activity in and around the thermal spring areas. Presence of helium can be accountable for disintegration of radioactive elements. The thermal springs specially Attri may be studied extensively for radioactive minerals at depth and for helium in future.

References

Acharya S (1966) The thermal springs of Orissa. In: The Explorer, pp. 29–35

Allen ET, Day AL (1935) Hot springs of Yellowstone National park. Carnegie Institute Washington. Publ No 466, p. 1

Barth TFW (1950) Volcanic geology, hot springs and geysers of Iceland. Carnegie Institute, Washington, publ. 587:1748

Chaterjee GC, Guha SK (1968) The problem of origin of high temperature springs of India. In: Proceeding Symposium (II), 23rd International Geology Congress, vol 17

Day AL (1939) Presidential address to Geological Society America on hot spring problem. Bull Geol Soc Am 50

Datta Munshi JS, Billgrami SK, Verma PK, Datta Munshi J, Yadav RN (1984) Gases from thermal springs of Bihar, India. National Academy "Science letters" India 7(9):291–293

Deb S, Mukherjee AL (1969) On the genesis of few groups of thermal springs in Chhotnagpur gneissic complex, India. J Geochem Soc India 4:1–9
Ellis AJ, Mahon WAJ (1977) Chemistry and geothermal systems. Academic press, New York
Ghosh PK (1954) Mineral springs of India. Rec Geol Surv India 80:541–558
Lindgren W (1935) Mineral deposits, 3rd edn. McGraw Hill, Newyork
Lovering TS, Morris HT (1965) Underground temperatures and heat flow in the east Tinctic district Utah. USGS, prof. Paper 504-F, pp. F1–F28
Ray P, Singh P, Acharya S (1990) A note on quartz pebble conglomerate around Sikheswar-Burhaparbat in Dhenkanal-Keonjhar district, Orissa. In: Proceeding of 77th Indian Science Congress Association, Cochin, p 33 (Abst)
Roy S, Rao RUM (1996) Regional heat flow and the perspective for the origin of hot springs in the Indian shield. GSI Spl Publ No-45:39–40
RaviShankar et al (1991) Geothermal atlas of India. GSI Spl Publ No-19
White DE (1957a) Magmatic, connate and metamorphic water. Geol Soc Am 68:1659–1682
White DE (1957b) Thermal waters of volcanic origin. Bull Geol Soc Am 68:1637–1658

Chapter 8
Uses of Geothermal Energy

Abstract This chapter deals with multipurpose applications of geothermal energy. The thermal spring is considered as source of non-conventional energy. The advantage of this energy source over other is that it is perennial, inexhaustible and pollution free. Researches for using this energy have been undertaken over decades. In fact reserve of minerals and fossil fuels are non-renewable and continuously diminishing in quantity. The day has been approaching- perhaps by the turn of the century- when the fuel energy is bound to confront the development of the nation. Man has been looking for harnessing other sources of energy. Geothermal source is catching the imagination of all because of least cost to the environment. The geothermal energy has been utilized for various purposes throughout the world. In India electricity is generated by utilizing geothermal energy at Manikaran. The low enthalpy geothermal energy resource has been utilized for various purposes other than generating electricity. The thermal springs of Odisha are categorized as low enthalpy type. Low temperature, low total dissolved solids (TDS) and low outflow of thermal springs of Odisha suggest various uses other than electricity generation. The utilization of hot water of the springs would be in hydrotherapy, specially for the treatment of rheumatic and arthritic conditions. People sometimes drink this water for the treatment of gout, gastric disorder, dyspepsia and high blood pressure etc. As the water contains some sulphur it would be useful for treatment of skin diseases such as acne, eczema etc. The medicinal value of these waters has been enhanced by the presence of radon content. The water in some cases is clean enough to be potable. The water is used for drinking and cooking purposes. Health resorts (spa) have been constructed near thermal springs. In the field of agriculture, space heating, food processing, geothermal energy has played a significant role. Considerable quantity of helium gas in the thermal springs would be utilized in nuclear and space research.

8.1 Introduction

Geothermal energy is perennial, pollution free and inexhaustible for human consumption. Possibilities of multipurpose economic utilization of geothermal energy for direct (non-electrical) and indirect (electrical) applications are being promoted

S. C. Mahala, *Geology, Chemistry and Genesis of Thermal Springs of Odisha, India*, SpringerBriefs in Earth Sciences, https://doi.org/10.1007/978-3-319-90002-5_8

in recent years. Attraction of geothermal energy is that it can be used at a little or no cost to the environment. Unlike fossil or fissionable fuels, it does not pollute the biosphere.

Till a couple of decades back, geothermal energy was not playing any significant role in the scenario of world energy production. However, lately geothermal energy scene is changing very fast with a rapid spurt in its direct and indirect use because of its eco-friendly, renewable and pollution free nature. Geothermal energy has been in use in some form or other since many centuries, though its planned development gathered momentum only after 1950s. Presently there are as many as twenty countries, which are utilizing this resource in varying degrees for various purposes.

8.2 History of Utilization of Geothermal Energy

Early humans probably used water of the hot springs for cooking, bathing and to keep warm. Recorded history sows its uses by Romans, Japanese, Turks, Icelanders, Central Europeans and the Maori of New Zealand for bathing, cooking and space heating. Baths in the Roman Empire, the middle kingdom of Chinese and the Turkish bath of Ottomans were some of the early uses of balneotherapy where body, health and hygiene were the social custom of the day. This custom has been extended to geothermal spas in Japan, Germany, Iceland, America and New Zealand. Early industrial applications include chemical extraction from the natural manifestations of steam, pools and mineral deposits in the Larderello region of Italy.

This indicates that people have used geothermal energy for various reasons at different period of times. The Romans used geothermally heated water in their bath-houses for centuries. They also used the water to treat illness and heat home. In Iceland and New Zealand, many people cooked their food by using this heat.

Detailed investigations in order to examine the possibility of economic multipurpose utilization of geothermal energy for direct and indirect applications are being promoted in recent years. At present about 80 countries of the world have this unique energy source but only a few countries like USA, Italy, New Zealand, Russia, Japan, China, Hungary, France and recently India are making use of it.

8.3 Technologies and Resource Types

Geothermal resources vary in temperature from 30–350 °C, and can either be dry, mainly steam, a mixture of steam and water or simply water. The temperature of the resource is a major determinant of the type of technologies required to extract the heat and the uses to which it can be put (Table 8.1). The following table lists the basic technologies normally utilized according to resource temperature (Chandrasekharam 2000).

Table 8.1 Basic technology commonly used (after Chandrasekharam 2000)

Reservoir temperature	Reservoir fluid	Common use	Technology commonly chosen
High Temperature >220 °C	Water or Steam	Power Generation Direct use	Flash & Binary cycle Direct fluid use, heat exchangers
Intermediate Temperature 100–200 °C	Water	Power Generation Direct use	Binary cycle Direct fluid use, heat exchangers
Low Temperature 50–150 °C	Water	Direct use	Direct fluid use, heat exchangers

8.4 Direct (Non-electrical) Use of Geothermal Energy

Direct or non-electrical utilization of geothermal energy refers to the immediate use of the heat energy rather than to its conversion to some other form. The primary forms of direct use include swimming, bathing, therapeutic use (Balneotherapy), space heating including district heating, agriculture mainly green house cultivation, aquaculture and industrial applications etc. (Lienau and Lund 1974; Lund 2001). Geothermal fluid temperatures required for direct use are generally lower than those for economic electric power generation. The Lindal diagram (Fig. 8.1) indicates the temperature range suitable for various direct use of geothermal energy resource. The agriculture and aquaculture requires the lowest temperatures i.e. from 25 to 90 °C. Space heating requires temperatures in the range of 40–100 °C. Basing on the temperature, cascading utilization of geothermal fluids for various applications can be done (Ramanmurty et al. 1996). Cascading utilization means using the energy first at highest temperature then passing the fluid through other processes at lower temperatures, thus, using the available thermal energy more efficiently. The diagram (Fig. 8.2) shows cascade utilization of geothermal energy.

8.4.1 *Balneological Use*

Since long people have used thermal spring water for bathing and for their health. Balneology, the practice of using natural mineral water for the treatment and cure of disease also has a long history.

From time immemorial, evidences are available that natural sources of hot springs have been used for curing chronic diseases. The Romans and Greeks were used to keep the thermal springs clean for drinking purposes. They were famous for their spa development. The word spa is used as a Latin abbreviation, which means "***Health through water***". People go to spas for improving health and appearance, getting away from stresses to refresh and revitalize the body and mind. A week's spa vacation provides necessary interlude to change one's pace of life and well-being.

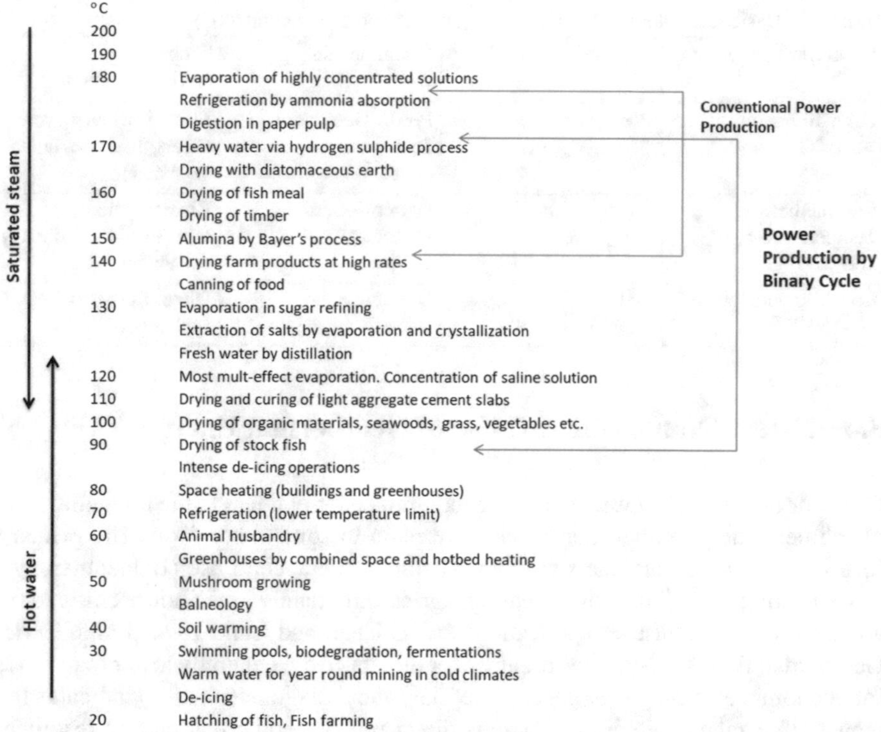

Fig. 8.1 Lindal diagram (approximate temperature requirements of geothermal fluids for various applications)

In ancient India the rulers and their associates used thermal water as a means of cure (Deb 1964). Most thermal spring waters contain enough soluble salts along with radon (radio-active element), which have imparted the therapeutic property to the water and, thus, the water is also called '***medicinal water***'.

High alkaline thermal spring water can be used safely for curing gastric disorders and dyspepsia associated with hyperacidity. The water is also taken internally for the treatment of gouts, liver disorders, high blood pressure etc. The water can be used for asthma and other ailments of respiratory tract. As the thermal spring generally contain sulphur, it would be useful for the treatment of chronic skin diseases like acne and eczema etc. Thus, thermal spring water is efficacious for various digestive disorders and is said to cure skin diseases.

8.4.2 Table Water

There has been an increasing demand of bottled thermal water in all the developed countries of the world. The thermal spring water is beneficial in correcting metabolic

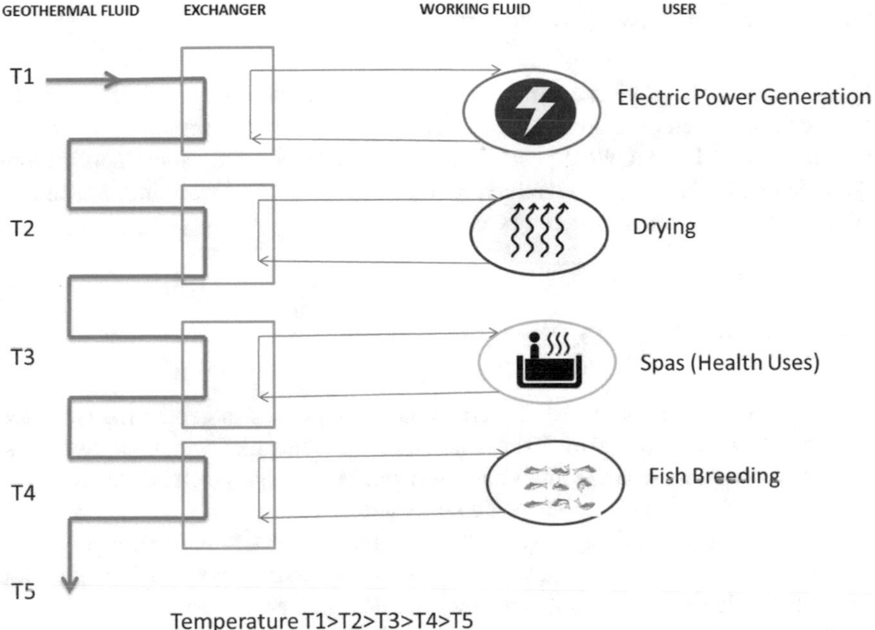

Fig. 8.2 Cascade utilization of geothermal fluid (After Mario Gaia et al. 1996)

disorders because of dissolve chemical constituents present in them. Development of table water for minor ailments is considered to be an important aspect of modern thermal water utilization. Ghosh (1954) opined that the water can be bottled at the site of the spring and sent to the people in all walks of life. The thermal spring water at Bakreswar (West Bengal) is being sold as mineral water.

8.4.3 Space Conditioning

Space conditioning includes both heating and cooling of apartment buildings. The natural geothermal heat energy is also an alternative to fossil fuel and firewood. Space heating with geothermal energy has wide spread applications. The most famous space-heating project in the world is the Reykjavik municipal heating project serving about 97% of the people in the capital city of Iceland. Buildings heated from individual wells are popular in Klamath Falls, Oregon and New Zealand (Lienau and Lund 1974).

8.4.4 Agribusiness Applications

Agribusiness applications are particularly attractive because they require heat at the lower end of the temperature range. Use of waste heat or the cascading of geothermal energy also has excellent possibilities. A number of agribusiness applications like green house heating, aquaculture, animal husbandry, soil warming, mushroom culture and biogas generation can be considered.

8.4.5 Green House Cultivation

Numerous commercially marketable crops have been raised in geothermally heated green houses all over the world. These include vegetables like cucumber, tomatoes, flowers, houseplants, tree seedlings and cacti etc. Many large green houses are operated as tropical gardens for sightseeing purposes.

Fish breeding and raising poultry through the use of geothermal energy have become successful enterprises now a day. A very successful prawn raising operation has been developed near Wairakei geothermal field in New Zealand.

Livestock raising facilities can encourage the growth of domestic animals by a controlled heating and cooling environment. Geothermal fluids can also be used for cleaning, sanitizing and drying of animal shelters and aquaculture.

8.4.6 Industrial Applications

Drying and dehydration at moderate-temperature are important uses of geothermal energy. Drying of various vegetables and fruits, and their processing can be done by the heat derived from thermal springs. The natural geothermal water can be used in the manufacture of oil from coconut, extraction of pulp from bamboo and drying of cashew nut etc. These industries need large quantities of hot water and thus a considerable amount of fossil fuel for heating can be saved.

The geothermal gases mainly Neon, Argon and Helium (rare gases) have wide applications in low temperature physics, space research and also in aeronautical and nuclear sciences. Besides, the strategic minerals like Borax and Fluorite associated with hot springs have important industrial applications.

8.5 Indirect Use—Power Generation from Geothermal Energy

High temperature geothermal reservoirs containing water and/or steam can provide steam to directly drive steam turbines for generation of electricity. More recently developed binary power plant technology enables more of the heat from the resource to be utilized for power generation. A combination of conventional flash and binary cycle technology is becoming increasingly popular. High temperature resources commonly produce either steam, or a mixture of steam and water. The steam and water are separated in a pressure vessel and the steam is piped to the power station where it drives turbine to generate electricity. The separated water is utilized in a binary cycle power plant to generate more power. A binary cycle power plant of 5 KW capacity was installed at Manikaran hot spring area (Vedantham 1996) to generate electric power.

8.6 Status of Use of Geothermal Energy in India

The government of India constituted a "Hot spring Committee" in 1966 with a view to examine the possibility of development of geothermal fields in the country for power generation and other direct uses (GSI 1996). The work done by the committee formed the basis of systematic geothermal exploration in the country. Geothermal exploration has been going on for more than four decades in the country and has been completed in Parvati valley and Beas valley (HP), Tapoban in Alakananda valley (UP), west coast (Maharastra), Tatapani and Salabardi geothermal field in Narmada-Tapi basin (MP) and Cambay basin of Gujrat (Fig. 8.3).

The geothermal resources in India are suitable mainly for direct applications with power generation at few areas. At present geothermal energy is mainly used for tourism by way of providing hot water baths at Basist in Beas valley and Manikaran in Parvati valley. Thermal water at Bakreshwar (West Bengal) is being sold as mineral water. Space heating of small huts is being carried out in Puga valley and Parvati valley with technical help from Regional Research Laboratory (RRL) Jammu, Central Electrical Authority (CEA) and Geological Survey of India (GSI). The tourism and forest huts are provided with space heating system in the Parvati valley (Vedantham 1996).

Since large amount of borax encrustation is available in Puga hot spring area, RRL installed a borax extraction plant to extract borax. Green house cultivation is being carried out in Chumathang in Ladakh and 41 varieties of plants including vegetables, fruits and flowers are grown. A cold storage plant of 7.5-ton capacity was installed at Manikaran by Indian Institute of Technology (IIT, Delhi) using geothermal energy. A binary cycle power plant of 5KW capacity was installed at Manikaran by National Aeronautic Laboratory (NAL), Bangalore for power generation. A project for mushroom cultivation and poultry farming at Puga (J&K) has been executed by RRL Jammu utilizing the geothermal energy (Vedantham 1996).

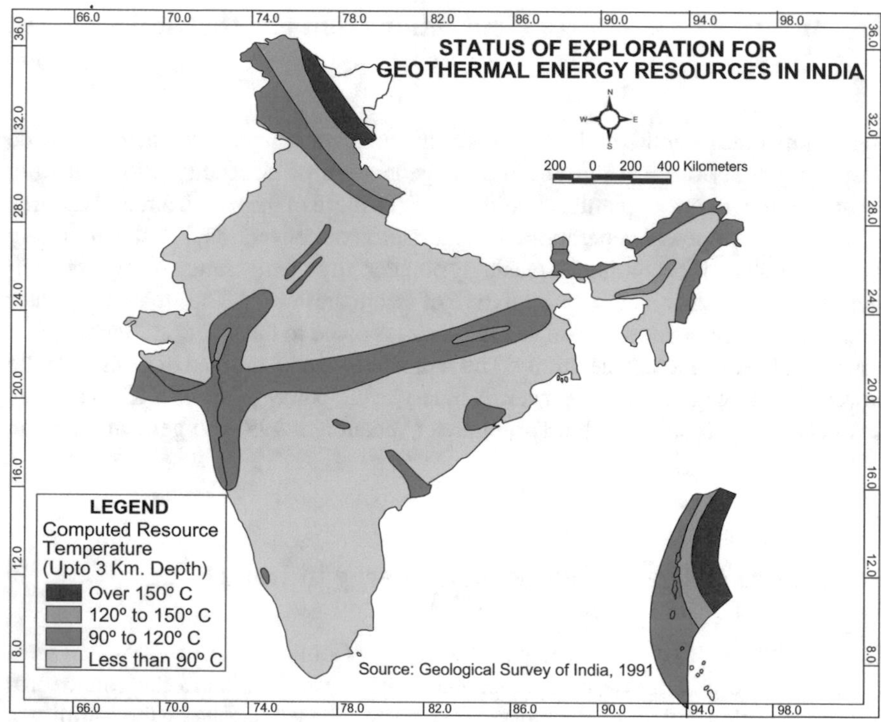

Fig. 8.3 Status of exploration for geothermal energy resources in India

8.7 Possible Uses of Existing Geothermal Resources in Odisha

Geothermal energy manifestations in form of thermal springs occur at eight places in Odisha. The geothermal energy resources of Odisha have a scope to be utilized in the fields of direct or non-electrical applications, as the thermal springs are of low enthalpy type.

Cashew processing and coconut oil extraction industries can be developed near the high temperature thermal springs of Attri, Tarabalo, Deuljhori and Badaberena as these industries consume huge quantity of hot water. Thus, instead of using firewood and fossil fuel, these industries can indirectly save large quantities of firewood, thereby protecting our forest resources and environment. The water of high discharge springs at Magarmuhan, Bankhol and Taptapani can be used locally for irrigation purposes. As already mentioned in chapter V (U.S.Salinity diagram, Fig. 5.24) the water of Magarmuhan, Bankhol, Taptapani, Boden and Badaberena thermal springs belongs to C_2S_1 class, and is good for soils of medium permeability for most plants. The water of Attri, Tarabalo and Deuljhori springs belongs to C_3 S_2 class that is satisfactory for plants having moderate tolerance.

Fig. 8.4 Photograph shows **a** Entrance Attri Tourist spot. **b** View of spring from entrance. **c** Hot water bath ponds. **d** Tourist rest houses

It is already mentioned that the thermal spring water is beneficial for the treatment of rheumatism, arthritis and skin diseases. Hence, spas (health resorts) can be developed near the thermal springs.

Government of Odisha has already developed tourist resorts at Attri and Taptapani (Fig. 8.4) with hot water bath facility. Similar health resorts and tourist spots can be developed at other thermal spring areas, which can boost the economy of the local people and can earn foreign exchange to the government exchequer. Besides, green house cultivation, fish breeding, mushroom culture and food preservation etc. can be done by using the geothermal energy resources (Lienau and Lund 1974). It has already been mentioned that the thermal springs issuing in Pre-Cambrian terrain yield radioactive elements. As a matter of fact, the radon content water can cure various ailments such as digestive disorders, skin diseases etc. The thermal spring water after proper examination can be bottled and sold as mineral water. Large storage tanks and water harvesting structures may be constructed in these areas to use the water locally for ground water recharge.

8.8 Discussion

Geothermal energy is a non-conventional one where cost, reliability and environmental advantages over conventional energy sources are quite important. It contributes both to energy supply, with electrical power generation and direct heat uses (Chandrasekharam 2000). The competing goals of increased energy production for worldwide social development and mitigation of release of atmosphere polluting gases are not compatible with today's fuel mix, which relies heavily on coal and petroleum. Development of geothermal energy has a large net positive impact on the environment as compared with conventional energy sources. This energy is practically perennial, pollution free, inexhaustible and eco-friendly. A holistic approach would certainly bring awareness about the importance of this energy and protection of its quality and quantity, would certainly help in conserving and managing these resources.

Though much has been said in regard to the positive aspects of geothermal energy it has its disadvantages. Geothermal energy is generally a highly localized resource and the processes used to extract energy move at a much higher rate than the processes that restore energy into the geothermal environment. In the same way that wind and hydropower rely upon certain wind speeds and certain levels of water, geothermal energy relies upon an area having certain level of thermal activity.

References

Chandrasekharam D (2000) Geothermal energy resources of India-Facts. In: Proceeding of geothermal power Asia 2000 conference, Manila, February 2000, pp 12–19

Deb S (1964) Investigation of thermal springs for the possibility of harnessing geothermal energy. Sci Cult 30(5):217–221

G.S.I (1996). Geothermal energy in India. Spl Publication No-45, p 391

Ghosh PK (1954) Mineral springs of India. Rec Geol Surv India 80:541–558

Lienau PJ, Lund JW (1974) Multipurpose use of Geothermal energy. Proceeding of the international conference on geothermal energy for industrial, agricultural and commercial-residential uses, Oregon Institute of Technology, Klamath falls, OR

Lund JW (2001) World status of geothermal energy use overview (1995–1999) (source Internet)

Gaia M, Sommaruga C, Zan L (1996) World geothermal resources map in relation to small electric power plants (binary cycle). GSI Spl Publ No 45, pp 119–125

Ramanmurty CV, Moon BR, Sharwan Kumar (1996) Cascaded utilization of geothermal energy at Puga valley geothermal field (Ladakh), India. GSI Spl Publ No-45, pp 127–136

Vedantham SS (1996) Power generation potential and non-electrical utilization of geothermal energy resources in India. GSI Spl Publ No-45, pp 7–10

Chapter 9
Environmental Issues

Abstract This chapter incorporates environment issues related to thermal springs. The pollution related to thermal springs is likely to be encountered due to discharge of certain gases and chemical constituents from them. The chemical data would serve as an invaluable base for detecting any contamination/pollution. It is mentioned in the literature that some toxic substances are being leached by natural process from the associated rocks and soils and pollute the thermal spring water. The ionic concentrations in the spring water are within the permissible limits. However, there is no threat of any disaster to be brought about by the gases released from these springs. Naturality should be allowed for these to protect the sanctity of the springs. A comparison study of emission of various gases and particulate matter from geothermal and conventional energy sources is attempted.

9.1 Introduction

The chemical interaction and toxic hazards associated with geothermal effluents must be considered before exploitation of geothermal resources. In his chapter attempt has been formulated to assess the possible environmental hazards of a geothermal field. Albeit in India, the exploitation of geothermal energy is yet to be undertaken intensively. In addition preventive measures have been suggested to mitigate the hazards related to thermal spring water and gases.

The quality of water is crucial to life. This must be taken into account the impact of the utilization of the water resources on the environment. Effective measures are suggested to maintain the quality of water and its management.

The main types of pollution, due to utilization of geothermal energy, likely to be encountered are atmospheric from the disposal of gas and water. The extent and detailed nature of problems varies for each field. Total discharge into the environment should also be considered because a continuous discharge having even lower concentration of toxic elements may be harmful as a result of cumulative effects (Pandey and Srivastav 1996). Concentration and toxicity effects of chemical constituents obtained in geothermal fluids are considered.

S. C. Mahala, *Geology, Chemistry and Genesis of Thermal Springs of Odisha, India*, SpringerBriefs in Earth Sciences, https://doi.org/10.1007/978-3-319-90002-5_9

9.2 Materials and Methods

Water and gas samples were collected from different thermal springs of Odisha. The water and gas samples were analyzed by standard laboratory methods. The summary of the results of chemical analyses of water and gas are given in (Tables 5.1 and 6.1) respectively.

9.3 Discussion

TDS values and concentration of cations and anions sometimes restrict the use of thermal spring water. Concentrations of Na, K, Ca, Mg (cations) and Cl, SO_4, CO_3 and HCO_3 (anions) in the thermal springs of Odisha are within the permissible limits as prescribed by WHO and ISO standards (Table 5.3). Chloride concentration of more than 300 mg/l is generally toxic to vegetation that results chlorosis, defoliation and burning of plants. However, it is below 300 mg/l in these springs that indicate no threat to the plants. Natural steams from thermal springs contain some non-condensable gases. These gases sometimes cause potential hazards to the surrounding environment. Ammonia being one of them is primarily an upper respiratory poison, which causes irritation of eyes and upper respiratory tract with coughing and vomiting. Though ammonia concentration in thermal spring gases of Odisha has not been determined the absence of such symptoms near the thermal springs indicates that there is no threat of this gas from the thermal springs. Hydrogen sulphide identifiable by its distinctive rotten egg smell, which is a noticeable geothermal effluent, considered as a pollutant to the atmosphere. But hydrogen sulphide once released, spreads into the air and eventually changes into SO_2 and sulphuric acid. Hydrogen sulphide emission is observed in thermal springs of Attri, Taptapani, Deuljhori and Tarabalo.

Development of geothermal energy has a large net positive impact on the environment as compared with conventional energy sources. When compared to fossil fuel energy sources such as coal and natural gas, geothermal energy produces almost near zero air emission. A comparison study of emission of nitrogen oxide, sulphur dioxide, carbon dioxide and particulate matter from the geothermal and conventional energy sources is given in the table below (Table 9.1).

The above table shows that the air emission from geothermal energy sources is very negligible in comparison to all other sources of conventional energy. A graphical presentation is shown in (Figs. 9.1, 9.2, 9.3 and 9.4).

In Odisha people put raw rice grains and coins into the pools of thermal springs out of religious belief. This causes unhygienic condition inside the pool and affects the quality of water of the springs. The author is aware of the fact that throwing coins into the river beds of the Ganges/Godavari is not a very uncommon practice in India and this is the outcome of religious belief to satisfy the Gods. For the huge water system of these rivers the pollutions caused need not be counted. In contrast however,

Table 9.1 Air emissions from different sources of energy

Sources of energy	Nitrogen oxides	Sulphur dioxides	Carbon dioxides	Particulate matters
Coal	4.31	10.39	2191	2.23
Oil	4	12	1677	–
Natural gas	2.96	0.22	1212	0.14
Geothermal (Flash)	0	0.35	60	0
Geothermal (Binary)	0	0	0	Negligible

N.B Units in lbs/megawatt hour
Source U.S Environmental Protection Agency, EPA2000

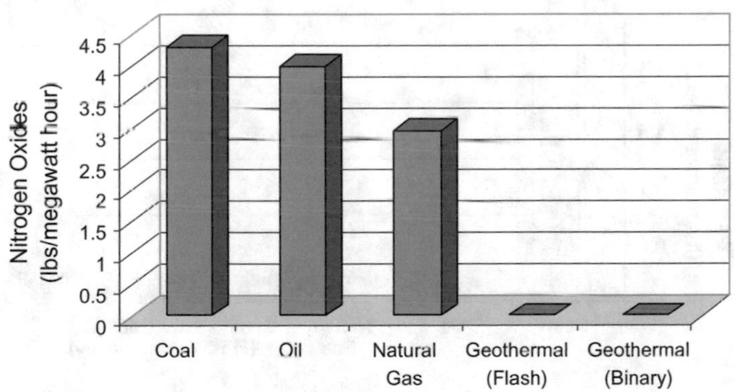

Fig. 9.1 Bar diagram showing nitrogen oxide emissions from different sources of energy

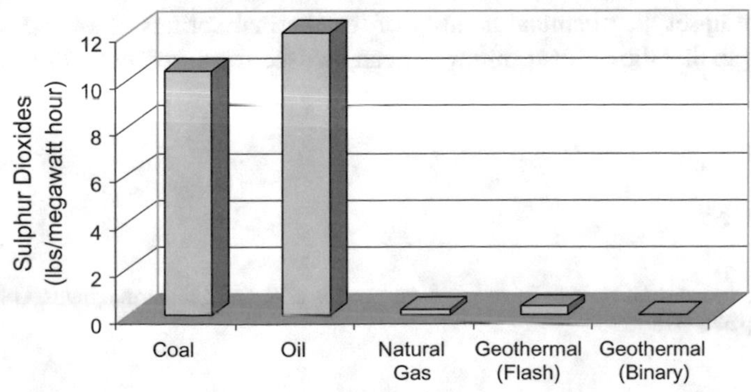

Fig. 9.2 Bar diagram showing sulphur dioxide emissions from different sources of energy

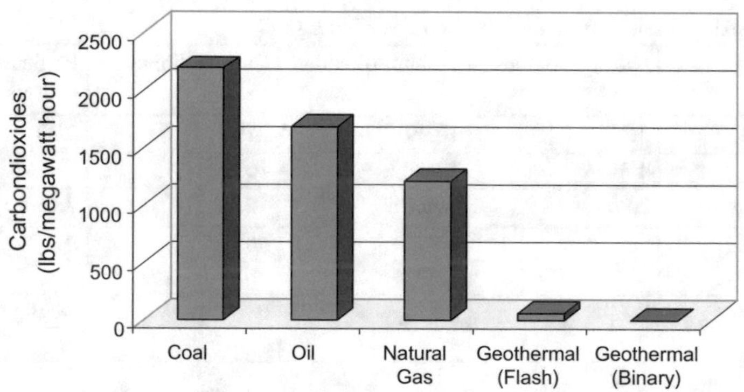

Fig. 9.3 Bar diagram showing carbon dioxide emissions from different sources of energy

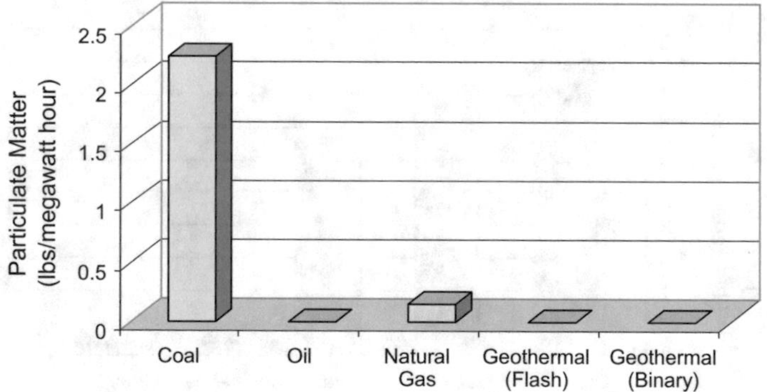

Fig. 9.4 Bar diagram showing particulate matter emissions from different sources of energy

throwing of coins into a very "restricted open system" like a thermal spring may in due course upset the chemical identity of the thermal springs. Awareness is to be developed so that the clean spring water can be used for different types of domestic purposes.

Reference

Pandey SN, Srivastav GC (1996) Environmental hazards of Indian geothermal fields. GSI Spl Publ No-45, pp 375–378

Chapter 10
Summary and Conclusion

Thermal springs are natural phenomena and are variously named as hot springs, geysers, fumaroles and mofetts due to their nature and surface manifestation. They emit heated water coming from deeper levels and are considered as the expression of earth's internal energy on the surface. The surface or meteoric water, penetrated to a great depth through structural breaks, get heated inside and subsequently come up to the surface in the form of hot springs. At places, heated steam/water coming out with force (like the old faithful) in form of fountains is termed as geyser.

Geothermal energy is the energy obtained by tapping the heat of the earth, usually from kilometers deep in the Earth's crust. This energy can be appraised as heat source of the earth. It is known that the temperature inside exceeds than at the surface of the earth. There exists a temperature gradient from the surface to the interior of the earth, as observed, averages about 1 °C for 30 m. The rate of temperature-increase with depth is known as the geothermal gradient (Lindgren 1935). The thermal energy stored in the earth's crust is manifested in the form of rising geothermal gradients (Valdiya 1987).

There are several weak tectonic zones and rifted grabens identified all over the country. More than 300 thermal springs are distributed along these weak tectonic zones (Ravi Shanker et al. 1991). It is reported that most of the hot springs are not related to magmatic activity except those of Himalayan geothermal province. Pandey and Srivastav (1996) mentioned that the hot springs are mainly of non-volcanic type. However, a few may be correlated to the volcanic source.

There is a crisis on the conventional energy resources because of rapid depletion, price rise and impact on the environment of the non-replenishable mineral fuels. It is, therefore, imperative to seek alternative renewable, non-polluting sources of energy especially to meet the widespread demand in the country as well as least cost to environment. Thermal springs can be considered as potential source that can be tapped for energy to cater the demand to some extent.

Till a couple of decade back, geothermal energy had an insignificant role in the scenario of world energy production. Lately, however, geothermal energy scene is changing very fast with a rapid spurt in its direct and indirect use, primarily due to eco-friendly, renewable and pollution free character. Recently high priority is being

S. C. Mahala, *Geology, Chemistry and Genesis of Thermal Springs of Odisha, India*, SpringerBriefs in Earth Sciences, https://doi.org/10.1007/978-3-319-90002-5_10

given to such non-conventional energy because of its environmentally benevolent nature (Lienau and Lund 1974; Valdiya 1987; Gawell and Bates 2004).

In geothermal systems, however, relatively deep penetration of fluids and steadily increasing temperature with depth facilitate changes in the original meteoric characters of the fluid. The descending fluid comes in contact with the litho-assemblages leaving their signatures on the fluid chemistry according to their relative susceptibility to chemical alteration processes. Water of these springs rich in mineral constituents and hence can be termed as "Mineral spring". The chemical constituents associated with thermal springs have impact on health of human being as well as of plants.

Geothermal activity in Odisha is manifested in the form of thermal springs at eight locations in different tectonic settings. Present study on different aspects of these springs reveals important findings on geological setting, hydrological character, chemical nature of water and gas and their utility. Regional geological mapping (Acharya 1966; Singh et al. 1995) indicates that the thermal springs of Odisha are mainly confined to crystalline schists and gneissic terrains of Precambrian period. The thermal springs are located within charnockite, khondalite, augen gneiss and mafic granulite belonging to Easternghats Supergroup or quartzite/quartz schists and metapelites of Iron Ore Supergroup. However, the spring at Boden is located within the Chhattisgarh basin of Proterozoic age and is related to Vindhyan Supergroup. There is no report of recent igneous activity associated with the thermal springs of Odisha. These springs emerge more or less along a line, which indicates a fault or fissure, through which heated water from deep regions comes quickly to the surface. The thermal springs of Odisha are pigeonholed under two categories of geothermal environment namely Mahanadi Valley geothermal province and Precambrian geothermal province.

The temperatures of these thermal springs vary from 32 to 67 °C. They are mildly alkaline in nature. The rate of discharge of thermal water from the springs varies from 33 to 84 lpm and the overall discharge rates remain the same throughout the year. The high temperature springs are characterized by steam and gas ebullitions and are marked with rings of waves on the water surface. Most of the springs are noted with sulphurous smell.

The chemistry of thermal water of Odisha can be grouped into two types namely (i) NaCl type and (ii) $NaHCO_3$ type except Boden thermal spring which is $CaHCO_3$ type. The total dissolved solids contents reveal that they are low in mineral content. The ratio of EC (Electrical conductivity) to TDS (Total dissolved solids) varies from 1.6 to 1.78. In cationic abundance Na is followed by Ca, Mg, and K. Sodium and potassium together constitute about 70–96% of the total cations in the thermal spring waters. The anion chemistry shows that chloride and bicarbonate are the dominant anions in the thermal spring water. As per US salinity diagram classification, the thermal spring water of Odisha fall in the category C_2S_1 and C_3S_2 class indicating their suitability for irrigation purposes.

The gaseous content of the thermal springs differ markedly from those of the atmosphere. The thermal springs of Odisha are categorized as nitrogen dominant type. The air saturated ground water is characterized by high nitrogen content. Comparable level of nitrogen in geothermal gases of Odisha suggests that the nitrogen has been derived from the atmosphere along with meteoric water. The rare gas helium particularly associated with Attri, Taptapani and Deuljhori might have formed due to radioactive decay going on inside the rocks. Methane may be produced as a result of heating and decay of organic matter present in the coal (Gondwana rocks).

The estimated base temperature from Na–K geothermometry at Tarabalo is the highest (277 °C) for Odisha and, therefore, this area deserves a priority for further exploration for assessing its geothermal energy potential.

The perennial nature of the thermal springs is because of availability of heat from within and water both from depth and surface. Hence, geothermal energy is classified as renewable. Use of this energy can at least partly conserve the wasting but polluting fossil fuel resources. The outlook for the use of geothermal energy depends on three factors:

(i) Demand for energy in general,
(ii) Inventory of available geothermal resources and
(iii) Competitive position of geothermal energy among other energy sources.

Demand for energy will naturally continue to grow everywhere. At the same time, there is growing global recognition of impacts of pollution due to energy production from fossil fuels and nuclear resources. The cost of geothermal fuel is predictable and is likely to remain stable. Experts predict that growth of uses of geothermal energy will reduce the atmospheric emissions of various polluting gases and particulate matter to avert global climatic change.

The springs in non-volcanic regions get heat from rise of geo-isotherms or penetration of water to greater depth. There is no evidence of recent magmatic activity around the thermal springs of Odisha. The rocks through which the water circulates at high temperature influence the mineral composition as well as radioactive character of the water. It is already mentioned that the Pre-cambrian rocks are carrier of radioactive elements. The thermal springs of Odisha are mostly issued in Pre-cambrian terrains. As the water of thermal springs has low TDS, the heat has been derived from a non-volcanic source. Thus, the normal heat flow due to metamorphism, decay of radioactive elements and exothermic reactions between mineral constituents are considered as the sources of heat for these springs.

Data recorded on geological, hydrological and chemical studies on these springs in Odisha sector are most promising for the multipurpose economic uses. The thermal spring water is being utilized for bathing and cooking by local people, who believe and drink this water for the treatment of gastric disorders and skin diseases. The water from springs particularly at Bankhol, Magarmuhan and Tarabalo is also used for agriculture purpose. Since the water flow is perennial, cultivation is done throughout the year around them. The heat related to the thermal spring has been used for hot water bath at Attri and Taptapani. There is a possibility for utilization of heat and water of the thermal springs for food processing and green house cultivation around these areas. Low temperature precludes the possibility of utilization of the water for production of electricity, as Puga valley condition does not exist here. The gases associated with thermal springs mainly helium, being the strategic element, can be extracted from Attri, Deuljhori and Taptapani thermal springs.

From the present findings the following conclusions may be drawn.

1. Eight numbers of thermal springs are located in different parts of Odisha.
2. The thermal springs are controlled by tectonic framework
   (i) Mahanadi graben,
   (ii) Along lineaments in Pre-Cambrian terrain.
3. Geologic and tectonic set up does not indicate the existence of any zone of volcanism, but Gondwana graben is caused naturally by vertical rifting.
4. Hydrological properties indicate that the thermal springs of Odisha belong to low enthalpy group (32 to 67 °C temperature), pH values show a neutral to mild alkaline nature and most of them are of sulphurous odour.
5. TDS values reveal that the springs are low in mineral content and are mainly of three types i.e. sodium chloride type, sodium bicarbonate type and calcium bicarbonate type.
6. Very high nitrogen, low oxygen, proportionately high helium and traces of methane, argon etc. is found in the liberated gas. Helium concentration at Attri is significant for exploitation by applying appropriate technology.
7. The thermal springs are considered to be of meteoric origin and the heat source may be derived from geothermal gradient, disintegration of radioactive elements or from the exothermic reactions during metamorphism.
8. The springs of Odisha can be utilized for direct applications in industries.
9. A conceptual model is drawn in order to decipher the origin of thermal springs all along the faults (Gondwana graben). It has also incorporated the possible source of water and heat responsible for the manifestation of thermal springs.

Exploitation of geothermal energy dates back to the first decade of the past century throughout the world. In India, exploitation is not at all intensive. More research and application in the field of geothermal energy source are essential in order to partially mitigate the shortfall of energy. Exploration and exploitation will further open up when programme of future energy requirement of the country is discussed.

References

Acharya S (1966) The thermal springs of Orissa. The Explorer, pp. 29–35

Gawel K, Bates D (2004) Geothermal literature assessment: Environmental issues Geothermal Energy Association

Lienau PJ, Lund JW (1974) Multipurpose use of geothermal energy. Proceedings of International conference on geothermal energy for industrial, agricultural and commercial residential uses, Oregon Institute of Technology, Klamath Falls, OR

Lindgren W (1935) Mineral deposits 3rd edition McGrawhill, New York

Pandey SN, Srivastav GC (1996) Environmental hazards of Indian geothermal fields. GSI Spl Publ No-45, pp 375–378

Ravi Shankar et al (1991) Geothermal Atlas of India. GSI Spl Publ No-19

Singh P, Das M, Ray P, Mahala S (1995) Thermal springs of Orissa: An appraisal. Utkal University, Spl Publ 1:244–254
Valdiya KS (1987) Environmental geology, Indian context. TATA McGrawhill publ p 582

Printed in the United States
By Bookmasters